Virtual
Organisms

Virtual Organisms

The Startling World of Artificial Life

MARK WARD

THOMAS DUNNE BOOKS

ST. MARTIN'S PRESS ⚏ NEW YORK

THOMAS DUNNE BOOKS.
An imprint of St. Martin's Press.

www.stmartins.com

ISBN 0-312-26691-X

First published in Great Britain by Macmillan,
an imprint of Macmillan Publishers Ltd

First U.S. Edition: November 2000

10 9 8 7 6 5 4 3 2 1

For Clare, always for Clare.

Preface

Men think about sex every ten seconds, or so popular mythology would have it. Well, I have had to think about sex a good deal more than that while writing this book, so it has not been all hard work.

However, the kind of sex I have been thinking about is not the kind most men's thoughts regularly turn to. Although sex between consenting bacteria or software programs excites few to passion it is, I hope, a fitting subject for a popular science book. And, I hope, an entertaining one too.

Popular science books usually fit into one of two categories. They are either centred on people or on ideas. Obviously in books about people the ideas do get a hearing and vice versa, but in general authors pick one or the other to act as the frame on which to hang the words.

Steven Levy's *Artificial Life*, one of the first books on the subject, was all about the people who did the pioneering work and established the field. Levy dwelt on the personalities and the route they took, both mental and physical, to get to where they are now. At the time Levy's book was written, 1992, the field of artificial life, or ALife, centred on these figures and it was their enthusiasm, often in the face of scepticism by their colleagues, that established the discipline in the first place. An emerging discipline needs mavericks to make its case and Levy's book did an admirable job of laying out the early history of the ALife movement.

Since the first ALife conference was held in Los Alamos in September 1987 the field is much more mature. Research in ALife today builds on the original work of these pioneers as much as it strikes out on its own. Because of this I feel that a round of interviews with the same pioneers would serve little useful purpose. So, in writing this book I have been more concerned with ideas than people and where ALife is going rather than where it has come from. All the history and people are still there, but I've tried to concentrate on how the field has been expanded and where it might be taking us.

There are some names in this book that Levy touched on. Now much more is known about them, so I have included details when I thought it was merited. There are many new names in these pages too. In the main they are people who were not doing ALife research in 1987 but who are now close to its heart.

Another reason for picking ideas rather than people is that it would be a formidable task to interview all those now involved in the ALife movement. This book is not an exhaustive treatment of every ALife project that is going on around the world. It could not hope to be, there is simply too much happening. That is more of a job for the first ALife encyclopaedia. What I have covered is the main ideas, the main practitioners and made a few gestures towards what the future of ALife might hold.

Making predictions is a tricky business at the best of times and one of the lessons I have learned while writing this book is that life is endlessly creative and can thwart even the most careful and tentative of predictions. For that reason I am reluctant to predict too precisely what fresh delights ALife will bring into being over the next few years. What I am sure about is that they will surprise us, shock us and challenge the most cherished assumptions we hold about ourselves. I, for one, can't wait.

No book is the work of only one person and I have to thank all the people who helped me while I was getting on with it. A huge thanks first to Chris Stewart without whom you would not be holding this book in your hands and I would have been unable to fulfil a long-held ambition. I owe you one, Chris. Also thanks to Catherine Whitaker and Tess Tattersall at Macmillan who whipped my woolly manuscript into shape. Then there are all the folk that I talked to while writing and researching. Thanks to Andrew Adamatzky, Rob Axtell, Mark Bedau, Randall Beer, Anna, Gavin and Katie Brooks, Rodney Brooks, Janet Bruten, Larry Burt, Peter Cochrane, Andy Coghlan, David Deamer, Peter Evans, Jose-Luis Fernández-Villacañas, Max Garzon, Richard Gregory, George Gumerman, Andrew Hodges, Owen Holland, Greg Hunt, Laurence Hurst, John Kumph, Chris Langton, Maja Mataric, Barry McMullin, Hans Moravec, Ian Morley, Andrew Pargellis, Tom Ray, Graeme Roeves, Peter Ross, Rudy Rucker, Ruud Schoonderwoerd, Gordon Selley, Tim Taylor, Adrian Thompson, Nicolas Walter, Chris Winter, and Stephen Wolfram. Thanks to those folks who looked over what I had written about them, corrected my mistakes and made it look like I know what I am talking about. Any errors that remain are of course unintentional and entirely my fault. Finally, the authors and publishers of quotations and extracts are acknowledged in the notes and references at the back of the book.

Preface

Thanks are also due to Clare, the most splendid person on the planet, who graciously put up with me neglecting her while I got this done.

Mark Ward
Surrey

Contents

Contents

Virtual Organisms

Introduction

For you must know that it is by one and the same ladder that nature descends to the production of things and the intellect ascends to the knowledge of them; and that the one and the other proceeds from unity, and returns to unity, passing through the multitude of things in the middle.

De la causa, principia e uno. Giordano Bruno[1]

The Universe is conspiring to kill you. But you shouldn't let it worry you too much. Seek reassurance in the fact that there is nothing personal about this. It's not a vendetta and it's not just you that it is working to extinguish. No, it's much worse than that. The Universe is ranging its awesome might against all the living organisms on our cosy, blue planet. Every louse, lover and leopard is going the same way.

Nor is this a destiny reserved for Earth and its inhabitants. It is shared by all the species on planets that we know nothing about. Our ignorance of them, and much of the life on Earth, may be total, but we all have one thing in common – our fate. We're doomed. There are no surprise endings to the greatest drama of them all.

The reason for this is physics. The first and second laws of thermodynamics to be exact. Between them they describe how things are and the direction they are heading.

The first law concerns itself with energy. It states baldly that there is a finite amount of energy in the Universe. This energy can be spread around and take different forms but there's only so much of it for you, me and the stars to use. The first law of thermodynamics is a physicist's way of saying there is no such thing as a free lunch.

One of the consequences of the law is that the amount of energy in a system stays the same. 'System' is shorthand for place where energy is being used. A steam-engine is a system and so is a waterfall. The Universe is the biggest system of them all.

Energy gets moved around, it passes from stars to plants to cows and in the process gets converted into different forms, but the total never changes. Where ever work is being done, a tree is being chopped down, an engine is powering a car or a flagellum on a bacteria is whirling, energy is being changed into different forms. All of which add up to the same amount. But while the total never changes the amount available to work with does. This is because whenever energy is being used to do work some of it – how much depends on the efficiency of the system in question – will escape into the wider world as heat.

This is where the second law comes in, it deals with heat and what happens to it. The law comes in a variety of guises one of which is sometimes known as the 'arrow of time'. It says the same thing you would if asked: What happens to hot things? They cool down. Not only does this law predict what happens to hot drinks, but also it applies to hot stars, galaxies, the Universe and everything in it. Everything is gradually running out of energy and wearing out.[2]

Very slowly, as bacteria, insects, humans, planets and galaxies go about their everyday existence, the energy they use is being redistributed. Part of this is to do with the fact that the Universe is expanding – the same amount of energy in a bigger space produces an overall cooling effect – but also some of this energy is being changed into a form – heat – that can't be used again. If there is less and less useable energy around then the Universe must be winding down. Which is why none of us are going to get out of this alive.

One crumb of comfort to be taken away from all this is that you will not have to worry about energy shortages any time soon. Our Sun has at least 6 billion years left in it before it starts to expand and get a little too close for comfort. The Universe will last even longer. So you have probably got time to finish reading this book and you have certainly got enough time to buy it.

Given that the Universe has a direction, you have to ask yourself how, in the face of such an uncaring and chaotic system, do livings things manage to resist all the forces that are constantly trying to do them down? If the whole Universe is winding down, what is keeping us alive? Although ultimately we will all fail in the attempt, briefly we do manage to assert ourselves even as the Universe rolls majestically towards its conclusion. How we do this is a difficult question. Answering it, finding out what life is and what keeps it living has kept generations of priests, philosophers, physicists and biologists amused for centuries. Lately though significant progress towards an answer has been made.

The Greek philosopher Aristotle thought that a vital force was responsible for giving life to organisms. For him this accounted for the special quality living things appeared to possess. He coined the term entelechy, by which he meant the force that gives function to form. Without it, said Aristotle, we would just be shapeless lumps of matter. With it, he says, we are vital, purposeful beings. The greatest advocates of this vitalist point of view were Henri Bergson and Hans Driesch. They died within months of each other in 1941 and to the end believed in entelechy and the vital force that kept the inner cogs whirling.

This vitalist view is waning largely because science is starting to make life look a little less special and in need of vital forces to animate it. The trend began with Copernicus, who rearranged the heavens to put humanity on the periphery rather than at the centre of the action. Now biology and artifical life are placing limits on where this special force can operate within the body.

If you remain unconvinced about the passions that overthrowing old ideas can arouse, you need only know the fate of Giordano Bruno, a quotation from whom opened this chapter. This rebel Dominican monk was among the first to take what Copernicus was saying seriously. He became something of a guerrilla theologist and, perhaps foolishly given the tenor of the times he lived in, spent his time travelling Europe spreading the word.

In his innocence Bruno took up what Copernicus was saying because he could not conceive of a God who had to operate within limits even though this is what orthodox Catholic beliefs asserted. This doctrine said that beyond Earth was a crystal sphere on which the stars were fixed. Beyond the sphere sat God.

Bruno thought it more fitting to believe in an infinite God who created an infinite Universe and many planets with life upon them. To Bruno this seemed a more God-like thing to accomplish than the creation of a crystal sphere. For his beliefs Bruno was tried as a heretic, then imprisoned and tortured. He was executed by being burned alive in 1600 on the Campo di Fiori in Rome. Before he was burnt at the stake he was gagged to stop him screaming heresy as he died.[3]

In 400 years we have come a long way. But in spite of the gradual accretion of knowledge, a lingering vitalism persists. Some modern-day scientists find it hard to let go of the idea that living things are not playing in a league of their own.

The view finds expression in some of the writings of scientists such as Karl Popper, Sir John Eccles, Roger Penrose and John Searle. Many others hold similar views. To one degree or another they think that there is something special about living things and, by implication, there is something extra special about humans. We appear after all to be the only organism to possess consciousness (whatever that is). Some were led to the view by their professional experiences. One such was Wilder Penfield, a psychologist who performed hundreds of neurosurgical operations on epileptic patients. This work convinced him that there was more to the brain than was revealed when he opened their skulls during surgery. He wrote: 'For my own part, I have come to the conclusion that it is simpler (and far easier to be logical) if one adopts the hypothesis that our being does consist of two fundamental elements.'[4] The elements he was thinking of were body and mind – the duality that has replaced the dead matter–vital force double act that Aristotle invoked.

These scientists are joined by Piers Paul Read, who wrote about humanity's cosmic position in the *Daily Mail* shortly after possible signs of extraterrestrial life were discovered in the Martian meteorite ALH84001. A moment that might, one would think, provoke sober reflection on our place in the Universe. Instead Read wrote: 'Quite possibly God, in the course of creation, created life forms in the Universe of which we are ignorant. Equally, long before the Big Bang He may have created beings of pure spirit. But he chose neither to take the form of animal or angel or a microbe on Mars. He chose to become a human being. Whether he likes it or not, man remains at the apex of creation.'[5]

Read's point is a variation on the theological argument from design. This says that the hand of a creator is revealed by the regular array of features found in the natural world. The proponents of this argument express incredulity when asked to accept that the diversity of the animal world has evolved. It is much easier to explain this variation, they say, if we accept that it was designed by God.

William Paley put forward the most famous conception of this argument. He asked what we would do if we were walking across a moor and found a watch lying on the path. We would assume that such an intricate device was the creation of a watchmaker. Given the intricate nature of many living organisms, he says, we must accept that these were designed too. Richard Dawkins, in his book *The Blind Watchmaker*, did a thorough job of showing just how it was entirely possible, and more believable, that the natural world evolved rather than was built.

It may be the case that living organisms were designed by God or whatever creator your religion demands. It would appear, however, that God is a very bad designer. You would expect an all-powerful entity to be better at creating creatures than the evidence on this planet suggests. Surely God would get everything right first time, yet there is evidence that many organs evolved separately many times. Eyes have evolved at least five times.[6]

Although our understanding is growing and the room for holding on to such ideas is shrinking, many still try and justify it by saying that what makes us special has simply not been discovered yet. Roger Penrose in *The Emperor's New Mind* invokes quantum effects to try and explain the mystery of consciousness, this being one of the few areas of physics that is stubbornly holding on to its secrets. It seems rather convenient that consciousness should reside in the place where it is hardest to get at.

This trend has led Chris Langton, one of the pioneers of the artificial life, or ALife, movement, to predict that when a computer can do everything scientists demand living things do we will be faced with the choice of either admitting that the computer process is alive or moving the goal posts to exclude the computer from the exclusive club of living organisms. Langton suspects the latter is more likely and he adds: 'Part (but only part) of the failure of AI to achieve "intelligence" is due to the fact that it is chasing a moving target. It used to be thought that playing chess required true intelligence. Now that computers can beat most humans at it, the ability to play chess is no longer considered an indicator of intelligence.'[7]

Hopefully this book will go some way towards showing why it is wrong to think that there is something special about life in general or humanity in particular. This is largely because artificial life studies are, *inter alia*, helping to show quite the opposite. No *élan vital* is at work within them, oiling the gears and keeping the whole mechanism ticking over. Everything you need to explain what is going on is right there.

This does not mean that life is unremarkable. ALife research and the fields associated with it are showing what a breathtaking gamble the whole enterprise is. Life remains special not because there are magical forces at work within it but because it manages to keep living without the need for supernatural aid. This only serves to make it all the more marvellous.

The continuum of living organisms stretches from viruses, which are little more than a stretch of DNA or RNA surrounded by a protein sheath, up to planetary ecosystems like Earth at the other. All of them use the same principles to – briefly – defy the first and second laws of thermodynamics and the attempts of the larger Universe to return them to the dust from whence they came.

Artificial Life research encompasses software simulations, robotics, protein electronics and even attempts to re-create the Earth's first living organisms. It is less concerned with what something is built of than with how it lives. It is concerned with dynamics and just how life keeps going.

All you, I and everything else need is information.

Life Rules

The second law of thermodynamics says that the Universe is winding down. It is gradually becoming more and more disordered. Energy in useful forms is slowly being dissipated from the places where it is used to do work out into the air or space in the form of heat where it can do no one any good. This disorder is measured by a concept called entropy and it is slowly increasing.

Systems that possess order – you, me, whirlwinds and whippoorwills – have a low entropy. We are suffered to remain in our living state because we are so good at creating disorder. We consume energy in a variety of forms but we excrete it largely as low-temperature, high-entropy heat.

We pay our debts to the second law, which strictly says we should not exist because there is a net decrease in entropy in a living system, by generating far more entropy and disorder than the small amount of order needed to keep our bodies functioning. This is the con trick played on the Universe by every living thing and so far we have not been caught out. With a bit of luck it will be a long time before we are. As scientist James Lovelock says: 'If your excretion of entropy is as large or larger than your internal generation of entropy, you will continue to live and remain a miraculous, improbable, but still legal avoidance of the second law of the Universe.'[8]

Information is the key to maintaining our inner peace and avoiding a descent into entropy. Built into all of our genes is a wonderful mechanism for preserving information – deoxyribonucleic acid (DNA). Genes have been characterized as selfish, but what they hold tightest of all is information about the body they are travelling around in. DNA and its little helper RNA are remarkable molecules. Both are strands made up of subunits called nucleotides. Attached to each nucleotide is one of four different bases. DNA has guanine, cytosine, adenine and thymine, RNA has the first three but uracil in place of thymine. A triplet of bases is known as a codon and, via DNA's faithful servant RNA, specifies the instructions for making one of 20 different amino acids. These link together in chains ranging from over 100 to 600 links to make up the proteins in every living thing on Earth. Proteins play a vital role in all living things. Some, such as collagen and silk, are used to support structures like cells and others, such as enzymes, catalyse important reactions that keep cells functioning. Proteins account for at least half the dry weight of many organisms. This simple system has given rise to the tens of millions of species living today.

FIRST BASE

		Uracil	Cytosine	Adenine	Guanine	
SECOND BASE	Uracil	phenylalanine	serine	tyrosine	cysteine	U
		phenylalanine	serine	tyrosine	cysteine	C
		leucine	serine	(stop sequence)	(stop sequence)	A
		leucine	serine	(stop sequence)	trytophan	G
	Cytosine	leucine	proline	histidine	arginine	U
		leucine	proline	histidine	arginine	C
		leucine	proline	glutamine	arginine	A
		leucine	proline	glutamine	arginine	G
	Adenine	isoleucine	threonine	asparagine	serine	U
		isoleucine	threonine	asparagine	serine	C
		isoleucine	threonine	lysine	arginine	A
		(start sequence)	threonine	lysine	arginine	G
	Guanine	valine	alanine	aspartic acid	glycine	U
		valine	alanine	aspartic acid	glycine	C
		valine	alanine	glutamic acid	glycine	A
		valine	alanine	glutamic acid	glycine	G

(Right-hand label: **THIRD BASE**)

How triplets of bases code for amino acids
e.g. UUU = phenylalanine

Written in the genes is all the information you need to make a new rabbit, rose or remora. Life is all about the preservation of this information. The whole grand plan centres on passing this data down the eons and making as few mistakes as possible in the copying. So far it is working really well.

Life keeps going using this information. The relationship between information and entropy was first discovered by Claude Shannon, an engineer who worked at Bell Telephone Laboratories in New Jersey during the 1940s and 50s. Shannon was investigating what prevents information being transmitted across a channel or telephone line. He found that fault lay with a hard-to-define quality that always seemed to be increasing whenever information was lost.[9] Shannon never witnessed a decrease in this quality in all the experiments he performed. Acting on the advice of mathematician John von Neumann, Shannon called this slippery quality entropy. Life is all about ensuring information is passed on, or transmitted, while all the time preventing entropy from corrupting the message. Life has found a way to ensure that entropy keeps increasing but not at the expense of its own survival or the integrity of the information it wants to transmit. Evidence of this can be found if we take another look at amino acids. The more abundant an amino acid is the more care is taken to ensure the information is read and preserved correctly. Six different codons produce leucine perhaps because it is so abundant.

This is the biggest and best trick that life has learned. Now ALife is helping us to understand just how it does it. It is starting to show that living things do not rely on the properties of chemicals to foster another generation, they depend on information encoded in chemical form. ALife is showing just how the dynamics of information can come to dominate over the properties of the materials living things are made of.[10]

ALife research has revealed that this shift occurs when a system is acting chaotically. Not chaotically in terms of 'badly organized' but chaos in the mathematical sense.

In the real world stability seems to reign. The seasons come and go, our hearts keep on beating and the taps keep dripping but in truth nothing is as safe or secure as it first appears. Living things cannot be captured by arid equations, they are too messy for that. Intuitively we feel that we have more in common with clouds than clocks and chaos helps explain why.

Predicting the movement and interaction of two bodies as they orbit the Sun is relatively straightforward. Newton comprehensively solved the mathematics of this problem. You could be forgiven for thinking the maths would not get much more complicated if a third body were added. In fact the addition of another object creates an intractable computation known as the n-body problem. The mutual interaction of the three bodies around a common gravitational centre, the Sun in the case of our Solar System, makes their exact movements impossible to predict. Only in the last decade have researchers been able to use chaos mathematics to confirm that this is the case. The implication is that the same is true of any system where multiple forces and influences are at work.

This interplay of objects and forces plays two roles. It keeps a system unpredictable, constantly threatening to upset the apparent equilibrium and plunge it into chaos, but at the same time the push-me-pull-you play of forces keeps it on the straight and narrow.

In complicated systems, such as dripping taps, air flows over wings and weather patterns, so many factors coincide that predicting their outcome over anything but the short term is, to all intents and purposes, impossible. The best that can be done is map out all possible outcomes and watch the subject move around the territory to get a feeling for what forces are at work and how they are weighed against each other. These 3-dimensional behavioural maps can be plotted on computer and are said to exist in phase space. This is an abstract concept more often used by physicists. You can think of it as a place where anything can happen. The constraints on what does happen are decided by what it is that you are modelling. If you want to model the movement of a pendulum, the dimensions of your phase space will be velocity, position and friction. The behaviour of the pendulum can be plotted at any moment as these forces act on it and it slowly loses momentum and comes to rest.

Plotting this behaviour in phase space can be revealing. It shows the relationships between the forces acting on a system and any patterns in that behaviour. Edward Lorenz, one of the pioneers of chaos theory, gave the name of 'the Lorenz attractor' to one of the patterns he discovered in phase space. The pattern resembled the face of an owl or a butterfly with its wings open. Other phase space patterns resemble closed fists with sharp drops abutting flat planes that, perhaps, represent a period of stability in the development of that system. Others are spiky like medieval weapons with sharp precipices between peaks.

What is becoming apparent is that life has to exist on a knife edge. Observation of fish populations and ant colonies, as well as experiments with artificial creatures, have shown that too much chaos produces a system that never gets a chance to settle down and develop. But if there is too little stimulation, everything stagnates. Between these extremes is a narrow, fertile region where just enough disturbance gets through to keep a heart beating healthily or a population thriving. Life needs a threat to keep it sharp. Entropy, or too much disorder, is one threat but the environment throws up others in the form of predators and climate changes.

This conception of life as a dynamic process – rather than an inherent property of the bits and pieces that living organisms are made of – has only recently begun to prevail. Prior to this change in perception life and living systems were thought by ecologists and biologists to be working towards a static state where environment and organism work in harmony. Now it is known that feedback and conflict is the norm, with species constantly wrestling for the upper hand and managing to survive and develop together as a result.

The approach of using biological metaphors and computer models of chaos to try and improve our understanding of natural and artificial systems is being used everywhere. Researchers at the Bionomics Institute in San Rafael, California, are starting to look at national economies as organisms or ecologies and have been surprised at how useful the analysis has been. Companies are starting to be compared to living, dynamical systems too and again the insights are piling up.

This method works because of the realization that the essential properties of life can be abstracted away from the bustling flora and fauna we find on Earth. In constructing computer models of living things we lose nothing because life is all about the preservation of an abstract property – information. Provided the behaviour that a computer model or a robot exhibits is lifelike then it is legitimate to use it to draw conclusions about the animals scuttling around on the planet. ALife pioneer Chris Langton says:

> The 'artificial' in Artificial Life refers to the component parts, not the emergent processes. If the component parts are implemented correctly, the processes they support are genuine – every bit as genuine as the natural processes they imitate.
>
> The big claim is that a properly organized set of artificial primitives carrying out the same functional roles as the biomolecules in natural living systems will support a process that will be 'alive' in the same way that natural organisms are alive. Artificial life will therefore be genuine life – it will simply be made of different stuff than the life that has evolved here on Earth.[11]

In many ways any artificially created life is much more useful than any of the real living creatures that are running around getting on with their lives. It is impossible to make a rabbit or beagle experience something for the first time again and again. They are too good at adapting to be fooled more than once. Their memory will help them cope. The same cannot be said about ALife creations; the memories of these creatures can be wiped time and again and their reactions recorded as if they were encountering an obstacle, a predator or a person for the first time.

ALife also helps us say more meaningful things about life in general. Biologists are doing a sterling job of capturing many of the properties of the creatures on Earth but it is hard for them to claim that what they are finding out has any universal relevance. We have no knowledge of life on other planets and how it goes about its daily life to compare with how ants, anteaters and antbears spend their days. ALife can help by creating new forms of life and watching what they do. In this way we can start to draw more meaningful conclusions about life because our theories draw on novel examples. ALife is as much about life-as-it-could-be as life-as-we-know-it.[12]

But before we can create new beings that share the properties of the life on Earth we need to know a lot more about current conceptions of living things – what they do and how they learned to do it. We need to draw up a checklist of the life all around us so we can be sure that our own novelties are doing the right things. To do that we have to begin where life started – at the bottom.

Chapter One

First Stirrings

During evolution there was great selection pressure for immediate action: crucial to our survival is the instant distinction of predator from prey and kin from foe, and the recognition of a potential mate. We cannot afford the delay of conscious thought or debate in the committees of the mind. We must compute the imperatives of recognition at the fastest speed and, therefore, in the earliest-evolved and unconscious recesses of the mind. This is why we all know intuitively what life is. It is edible, lovable or lethal.

The Ages of Gaia, James Lovelock[1]

Peer Pressure

Physicists are among the luckiest of scientists because they can see almost to the edge of creation. In contrast the view that biologists have of their own big bang, the origins of life, is much muddier and indistinct.

Theories about events during the first moments of the Universe start a mere 10^{-43} seconds after the Big Bang. The difference between zero and 10^{-43} seconds later might not sound like enough to matter. In fact its very nearly nothing at all, although the moment 10^{-43} seconds after the Big Bang has a name, the Planck time. The early Universe managed to get a lot done in a tiny amount of time. It went from being infinitely hot to a temperature of only 10^{28} Celsius by 10^{-35} seconds after the Big Bang. It has cooled down a lot since then. Today the average temperature of space is a chilly −270°C. In a tiny fraction of a second the Universe expanded from nearly nothing into something as big as a pea. A very heavy, very hot pea.

Physicists are not entirely sure what was happening before 10^{-43} seconds but they have their suspicions. They think that before this time all the familiar forces of today (electromagnetic, gravity, the weak and strong nuclear forces) were united into one superforce. Gravity dropped out at 10^{-43} seconds and by 10^{-35} seconds the others had separated out as well. The interplay of these forces sets the Universe into the form we see it today by restricting the ways that subatomic particles can interact. But 15 billion years ago they were only just getting started as the Universe cooled and expanded enough for them to separate out. Testable theories about when and how three of the forces mix and match are well established, but it is proving difficult for physicists to work out how gravity fits in with the others. Fitting it in and getting closer to the Big Bang requires a quantum theory of gravity, something that has so far proved elusive.[2]

So if you want to look the creator, or creation, in the face, become a physicist, not a biologist. In comparison to the pin-sharp hindsight of physicists, biologists are cursed with myopia. When they look back they see only shapes and suggestions in soil and rocks not certainties written across the skies. Biologists have yet to agree when and how life got going, let alone plot the order of events since.

Cheap Meat

Earth is around 4.6 billion years old, over a quarter of the age of the Universe. This is an unimaginably huge span of time and one that is hard to get to grips with. My mental ready reckoner can cope with life spans of about a hundred years and histories of civilizations a couple of thousand years old but it cuts off well before ten thousand years, never mind one hundred thousand times as long as that.

An analogy may make it slightly easier to appreciate. If we take a mile to represent a million years and you were to drive at a steady 60 mph, it would take three days, four hours and 36 minutes of non-stop driving to journey from the Earth's beginning to the present day. Four thousand six hundred miles is roughly the distance by road between Lisbon in Portugal and Athens, Greece. It's a little further than the distance by air between London and Anchorage, Alaska.[3]

It's not just you and I that have problems handling these great swathes of time. To make them easier to deal with scientists who work with geological time scales on a daily basis have divided up the span of time between then and now into a series of eons, eras, periods and epochs. The eons are the largest divisions and the others are sub-divisions within these.

GEOLOGICAL TIME SCALE

When eon started	Eons	Eras	Periods	Epochs
4600	Hadean			
3800	Archean			
2500	Proterozoic			
580	Phanerozoic	Palaeozoic 580-245	Cambrian 580-500 Ordovician 500-440 Silurian 440-400 Devonian 400-345 Carboniferous 345-290 Permian 290-245	
		Mesozoic 245-66	Triassic 245-195 Jurassic 195-138 Cretaceous 138-66	
		Cenozoic 66-0	Palaeogene	Palaeocene 66-54 Eocene 54-38 Oligocene 38-26
Now			Neogene	Miocene 26-7 Pleiocene 7-2 Pleistocene 2-0.1 Recent 0.1-now

At first glance this division seems arbitrary and devoid of sense. There is no set span of time between eons, they do not fall regularly every billion years as any sensible dating system would have them do. Instead the earliest eon – the Hadean – lasted 600 million years but that which followed it – the Archaean – lasted over twice as long. The Hadean encompasses the time from the formation of the Earth to the age of the oldest rocks.

Memorizing this table is a formidable task. Harvard geology and zoology professor Stephen Jay Gould used to lighten the load by turning the task into a mnemonics contest. He says that the 'all-time champion' turned the list of eras and periods into a review of a pornographic movie called Cheap Meat.[4]

But there is a subtle intelligence at work in this chronology. The divisions are not arbitrary, they reflect the sequence in which the rocks of the Earth were laid down, changes in fossil forms and even great events that affected living organisms. The boundaries between the two most recent eras mark great extinctions, one of which did for the dinosaurs. The break between the Proterozoic and Palaeozoic eras around 570 million years ago marks the great blooming of life known as the Cambrian Explosion. Using the 1 mile:1 million years scale this is roughly the distance between London and Aberdeen. However, Athens, Anchorage and the origins of life lie much further afield.

Hell on Earth

The earliest eon of Earth's history is known as the Hadean and it is aptly named because the early Earth was a hellish place. Hades is the name for the Greek kingdom of the dead and the description fits because at this time Earth was devoid of life. It was little more than an intensely radioactive molten-lava fireball. There was no land, lakes or seas, but the planet was brimming with the chemicals that would soon spring into life.

While there is agreement about the state of the planet during the early centuries of the Hadean eon, there is no such accord about what happened next. The Hadean was the place that life started and it has proved just as fertile a ground for theories about the origins of life. Every stage in the story of life is fiercely contested. Arguments rage about what chemicals were present and in what concentrations, what reactions they might have taken part in and the energy sources they might have tapped. There is even dispute over where these chemicals might have come from in the first place.

There are two reasons why the origins of life are so hotly disputed. Firstly there is no hard evidence to study, no fossils to point at and poke, and secondly no one has worked out how exactly you get from a seething chemical broth to self-replicating, stable chemical forms to cells and then on to you and me.

There are no fossils because any dead proto-cells that were around at this time would have been destroyed by geological processes over 4 billion years or so since they came into being. The oldest recognizable microfossils are around 3.5 billion years old, but life itself is thought to be older, perhaps half a billion years older.[5]

Palaeobiologists are not on a wild goose chase when they go looking for these traces. There was definitely something living all those years ago, there is more than enough second-hand evidence for that. You might think that a slowly cooling fireball the size of the Earth would make it difficult, if not impossible, for life to begin. Yet this is exactly what happened. In fact life is thought to have got started several times only to be snuffed out by the impact of a large meteor, comet or moonlet.

Rock of Ages

The Isua rocks of western Greenland and those from the nearby Akilia island are the most ancient rocks on Earth at around 3.8 billion years old.[6] They mark the end of the Hadean eon and they bear the unmistakable stamp of life.

In 1988 Manfred Schidlowski[7] and his colleagues reported their discovery that the ratio of carbon-12 and carbon-13 isotopes in the Isua rocks was not as it should be. Usually these isotopes are found in rocks in roughly the same proportions. Yet in the Isua rocks the carbon-12 isotope predominates. The only process that uses the slightly lighter carbon-12 and therefore might lead to it being found in greater quantities in very old sedimentary rocks is photosynthesis – the process of using light energy to make carbohydrates.[8]

The Akilia rocks bear different signs of the same process. These rocks contain patterns of minerals known as banded iron formations (BIFs). The thin bands in these formations are made up of iron oxides – haematite and magnatite – formed when oxygen combined with iron ions in the early oceans.[9] The atmosphere of Earth in this eon was dominated by carbon dioxide so it has been claimed that the mineral layers formed because of the waste oxygen gases given off by photosynthetic bacteria.

If you were driving that car through history, you would do well to keep the windows rolled up when passing through this eon. There is no breathable oxygen in the atmosphere, instead it is made up of carbon dioxide, formaldehyde and hydrogen cyanide. Until 1994 this last gas was used by some US States in executions. The only other things in the atmosphere are ammonia, hydrogen sulphide and methane. All of which are superheated by the volcanic and nuclear activity of the newborn planet.

The picture is emerging that a mere 600 million years after the Earth was born life had taken hold. It got started nearly as soon as the Earth had cooled enough to produce a cracked crust riddled with niches for life to live in. The only question that remains to be answered now is: How?

Chemical Chance

Embarrassingly for biologists their success at explaining evolution and the transmission of genetic characteristics has not been matched by a similar level of success regarding theories about the origins of life. As ALife researcher Walter Fontana puts it: 'Biology has claim to two theories unto itself: Darwin's natural selection and Mendel's transmission rules. Both are correct, their joint operation can be nicely formalized, and together they are insufficient to account for the history of life as we know it.'[10]

The problem is that these theories deal with living organisms, but they make no suggestions about how these organisms may have arisen in the first place. Many, many theories have been put forward to explain the origins of life. Some are more plausible than others. The most respectable start with the seething stew of chemicals that was present on the surface of the Earth a little over four billion years ago.

As the Hadean eon ended and gave way to the Archaean the planet began to cool. The water vapour that until then was kept in the upper atmosphere by the heat of the Earth's surface fell, condensed and became rain. The rain fell for centuries. There was so much rain that warm shallow seas were created. The falling water eroded the young rocks and the run-off formed chemical-rich rivers and streams that mixed vital nutrients and delivered them to the new seas.

Soon the seas were swimming with chemicals. The cycle of light and dark, heat and cold, evaporation and inundation turned the world into a vast vat, providing energy and opportunity for compounds to mix and match. As this global mixing went on, the key chemical trends of self-reference and auto-catalysis became established. These are important because they help chemical structures maintain themselves in self-sustaining cycles. Catalysts are like spanners and can be used again and again without themselves being changed, they facilitate a reaction but are not changed by it and are free to do the same thing again and again. A short description of a living being might be a self-sustaining entity. This could apply to a man, a mouse, an amoeba or a mass of chemicals in a tidal pool. Lynn Margulis, one of the most respected researchers into early life, says: 'Today, although all of the chemicals in our bodies are continually replaced, we do not change our names or think of ourselves as different because of it. Our organization is preserved, or rather it preserves itself.'[11] Margulis sees common forces at work at every level. She claims that molecular complexes that arise spontaneously are doing the same thing as clusters of RNA that collaborate to create amino acids and other biogenic chemicals. Both of these share characteristics with crude cells that replenish themselves from their surroundings. In this commonality of action she says: '. . . we begin to see the winding road that self-organizing structures travelled on their journey toward the living cell.'[12]

Artificial Life researchers are helping to show that getting started on this journey is relatively straightforward. Theoretical biologist Stuart Kauffman, who works at the Santa Fe Institute, has built computer models of what happens when lots of simple polymers are mixed together. Polymers, such as DNA, are long chain molecules with many repeating sub-units. Kauffman started with soups of short polymers and gave them the ability to take part in different catalytic reactions, including those that make more of the short chains. The molecules and compounds bump around making and breaking chains and forming new longer polymers as well as more of the ones that were there first. There comes a point when all of the catalysts needed to maintain the different polymers are present and the system becomes self-sustaining. Out of the random and chaotic mixing of chemicals emerges order and self-maintaining systems. The sparks of life.

It should be stressed that Kauffman's chemistry is, initially at least, all about self-maintenance rather than self-replication. It tries to show how stable compounds can arise rather than how they can reproduce. However, self-maintenance is clearly a long way down the path of self-replication and it may have been that out of a stable population of such chemicals self-replication emerged.

Kauffman claims that these 'auto-catalytic sets' as he calls them show that emergent order is fundamental and arises essentially for free. We do not need to appeal to luck and probability to explain how a sea of blindly reacting chemicals gave rise to life. Which is just as well, because as Sir Fred Hoyle said, there is more chance that a hurricane passing through a scrapyard will assemble a working 747 aircraft than random collisions in the early oceans would eventually produce you and me. Or as planetary biologist James Lovelock has put it: '. . . life is characterized by an omnipresence of improbability that would make winning a sweepstake every day for a year seem trivial by comparison.'[13]

But once you admit that certain collections of chemicals can boot-strap themselves into ordered self-sustaining states the emergence of life becomes less problematic. As Kauffman says: 'Spontaneous order has been as potent as natural selection in the creation of the living world.'[14]

This does not mean that the problems have been solved. Showing how it *can* be done is no substitute for finding out *how* it is done. Kauffman's work, though suggestive, has all been with computer models, but now he is starting to mix chemicals to see what emerges and whether auto-catalytic structures emerge.

Kauffman was not the first to advance the idea that some populations of chemicals are self-sustaining. Over twenty-five years ago Humberto Maturana and Francisco Varela introduced the idea of autopoiesis in an attempt to capture the self-preserving properties of living systems. Autopoietic systems are self-sustaining, they take the materials they need from the world around them and turn them to their own purposes. As an epithet it can be applied to everything from the first cells right up to the largest animals. Typically though it is used to describe basic organisms such as pro-karyotic bacteria. This is because those interested in auto-poiesis prefer to study it raw, free of the complications that studying a larger animal would bring.

Maturana and Varela were among the first to test their ideas using a computer model. In 1997 Barry McMullin, an ALife researcher from Dublin City University, revisited their work and created an artificial chemistry to test their theories. McMullin did this partly for historical reasons and partly so that other ALife researchers could play with the earlier system or use it as a test bed to try out their own theories.

His artificial chemistry was very simple and had only substrates, catalysts and links. Using these few elements McMullin wanted to see if any kind of autopoietic system could arise, if order would emerge spontaneously. In any one run of the system McMullin seeded a space with a few of his chemicals and let it run. When he reported his first results McMullin had not had much success but he was persevering and had come tantalisingly close on a couple of occasions to seeing an enclosed cell spontaneously form.[15] His work continues.

His is not the only artificial chemistry that ALife researchers are playing with. The previously mentioned Walter Fontana from the Santa Fe Institute and Leo Buss from the Department of Biology at Yale are also working in this field to see just how much order you can get for free. Back in the real world, however, no one knows which chemicals are needed and in what order they have to react to create the structures we find in the cells of every living thing. One way to find out might be to try and re-create the conditions at the end of the Hadean eon.

In 1953 Harold Urey and Stanley Miller did just this. In a famous series of experiments the pair tried to produce some of the chemical building blocks of life. Their model world was two large laboratory flasks joined by tubing. One flask contained water to represent the early ocean and the other flask contained hydrogen, ammonia and methane, the gases thought, at that time, to be present during Archaean times. The water was boiled and its vapour was mixed with the other gases. The mixture was regularly zapped with electricity to simulate ancient lightning.

At the end of the week they cracked open the flask holding the water and inspected the contents. They found lots of organic compounds, including two amino acids that are only produced by living organisms. In later experiments they, and others who have reproduced their work, managed to produce the fatty acids found in cell membranes and even chemical complexes that were the building blocks of DNA and RNA.

They thought they had found the origins of life and the mechanism for creating the delicate chemicals needed for life. It wasn't at all difficult to create the organic compounds. Sadly, the promise of this experiment has not been borne out. For one thing, the mix of gases they used is now thought to be wrong. It is now widely believed that the primordial atmosphere was largely made up of carbon dioxide and nitrogen with far smaller concentrations of the ammonia, methane and hydrogen compounds Miller used. This mix is much less reactive than that which Miller and Urey used and is far harder to force into biotic compounds.[16]

Others have criticized Miller's work because the substances he obtained are chemically very reactive and tend to get broken down quickly by other chemicals. Only if carefully preserved do these compounds persist. As James Orgel and others have pointed out, there was no one around 3.5 billion years ago to preserve the compounds and carefully mix them together.

Don't Start

This is the big problem that has yet to be solved. How did the highly reactive compounds found inside all living things manage to survive and perpetuate themselves in the primordial sea? If RNA or DNA is removed from its protective cell it quickly degrades into shorter chunks.

In 1967 Speigelman[17] showed that if you put sections of replicating molecules in a test tube and supply them with nutrients they will continue to replicate. However, the replicating chains gradually get smaller and errors start to creep in. In this surrogate tidal pool evolution went towards greater simplicity. The same is true of many other organic compounds that only remain intact when inside a cell. Put them naked into water or a liquid full of ions like the sea and they swiftly break down into their component parts. But if these chemicals cannot form spontaneously in the sea where can they form? Outer space?

Actually that is not as crazy an idea as it sounds. During the Hadean and Archaean eons Earth was regularly being bombarded by comets, meteors and asteroids. To get an idea of the size of some of these objects just glance at the Moon. It is thought that a planet the size of Mars, or possibly bigger, crashed into the Earth, freeing enough debris to form its satellite.[18] While impacts on the scale of the one that formed the Moon were mercifully rare other smaller impacts were not. At this time the Earth was regularly being pummelled by celestial infall. Some of it may have been the rocks thrown up by the impact that created the Moon.

Complex chemical compounds may have been hitching a

ride on one of these rocks. Organic compounds have been spotted in outer space[19] and certainly carbon was one of the main ingredients in the rocks that were smashing into the planet.

It has been estimated that if all the carbon delivered by this celestial infall over a 100 million-year period was collected together it would form a layer thick enough to cover the Earth to a depth of 20 centimetres.[20] It is just as well that this carbon was delivered when it was because it is important for almost all of the chemical compounds found in living things. In the average human body there is enough carbon to make a 10-kilogram bag of coke. Carbon is found in sugars, one of which forms the backbone of DNA, without which none of us would be here. It is also essential for making carbohydrates, which serve as energy stores in plants and form the shells of many insects.

The theory that life began off Earth got a boost in August 1996 when NASA scientists reported that they had found biotic compounds in a meteorite recovered from the Allan Hills region of Antarctica. Meteorite ALH84001 was thought to be 4.5 billion years old and at the time was taken as proof that life had existed on Mars and that it could have been delivered to Earth from outer space. Not on ALH84001 perhaps but another rock from elsewhere in the Solar System. However, further analysis of the chemicals trapped within the Martian rock have thrown doubt on the earlier claims that they were evidence of life on other planets.[21]

But if ALH84001 is not evidence enough, and it is looking increasingly like it is not, more compelling clues can be found in older, larger rocks. In 1969 a meteorite was unearthed in Murchison, Australia. It was collected shortly after it fell to Earth and quickly subjected to analysis. It was found to be one of the most pristine of a class of meteorites known as carbonaceous chondrites – this just means that they are rich in organic compounds and contain some distinctive features. The Murchison meteorite is believed to be a fragment of a comet because it has a relatively high water content – comets are thought to be large balls of dirty ice. The analysis also revealed that the meteorite harboured many biotic compounds. In particular it was home to a collection of amino acids – the building blocks of proteins and essential for life on Earth. Nineteen of these amino acids are the familiar ones found in all living things on Earth, many more were made up of combinations never found in terrestrial organisms.

What is clear from the analysis of this meteorite is that complex organic compounds can form in interstellar material. Given that comets are regular visitors to our Solar system it is reasonable to think that they might have left something behind when they passed by. Something that perhaps landed in an ocean and got life going. Some astrobiologists believe that up to a quarter of Halley's Comet is made up of organic molecules and other comets may well have similar proportions of biotic compounds on board. Even if they don't there are other ways that such important chemicals could be delivered. Interplanetary dust is constantly falling to Earth after it has been floating around in space or the atmosphere for years.[22]

Other experiments have shown that biotic compounds can survive in space. The fate of a NASA satellite known as the Long Duration Exposure Facility has bolstered the case for the argument that the seeds of life arrived from off-planet.

The LDEF was put into orbit to act like a punch bag and

soak up everything outer space could throw at it. It spent nearly six years in orbit and was retrieved by the space shuttle *Columbia* in 1990. On board it were experiments to measure micro-meteorite impacts, radiation and cosmic rays. Also flying on the satellite were some bacteria being tested to see how long they could last in the harsh vacuum of space. Astonishingly, the bugs survived, showing that it is not impossible that extra-terrestrial material might have given life on Earth its start.[23]

In 1995 astronomers reported the discovery of the amino acid glycine in the giant dust cloud at the centre of our galaxy in the constellation of Sagittarius. Glycine is the simplest of the amino acids and is a component of many proteins. It was not the first time that such chemicals have been detected in space. Ammonia was the first to be spotted, in 1968, and since then nearly 100 complex molecules ranging from alcohol to formaldehyde have been discovered in deep space.

Critics of the off-Earth theory have pointed out that even if all the asteroids, comets and meteorites that crashed into Earth were chock full of organic molecules they still did not arrive in large enough quantities to get life going. Once diluted in the oceans the resulting soup would be a very weak brew incapable of sustaining life. The real primordial soup was probably much thicker. NASA is planning missions to Saturn's satellite Titan and Europa, one of Jupiter's satellites, because it is thought that conditions on these moons resemble those found on the Archaean Earth. Europa is a candidate because NASA astrobiologists think that beneath its icy surface are oceans warm enough to support life. The space agency is interested in Titan because conditions there may be similar to those of early Earth. It may even have oceans on its surface, even though they are likely to be methane seas rather than water. The surface of Titan is thought to be a chilly −180°C but NASA is still hoping to find evidence of life.

But as the critics of the extra-terrestrial origins theory point out, if living things can survive in the harsh vacuum of space, regularly bombarded by high-energy cosmic rays and ultraviolet light, surely a more benign place, somewhere that was warm, wet and rich in raw materials, such as the infant Earth, would be an even better place for it to get started?

Bugs Like it Hot

The Archaean seas are the leading candidates for the cradle of life. But it is not thought that life began in the body of the ocean, rather the belief is that life got going on the margins, on the beaches or around the hydrothermal vents on the sea floor.

Hydrothermal vents are cracks in the ocean floor through which pour scalding gases and water. The cocktail of gases streaming out of these holes can reach 350°C or more but this does not discourage life from colonizing these places. However inhospitable conditions seem around these vents they are a pocket paradise compared to what can be found on the rest of the sea floor. These vents with their lava flows and steaming cracks are surrounded by weird ecologies of giant worms, albino crabs and shrimps. The bigger organisms stay alive by feeding on the micro-organisms sitting at the bottom of the food chain in this strange habitat: bacteria. These bacteria are examples of extremophiles, micro-organisms that flourish under very harsh conditions. These bacteria are remarkable not just because they manage to survive around the hydrothermal vents but because of the way they stay alive. They synthesize food from hydrogen sulphide and carbon dioxide in the same way terrestrial plants use carbon dioxide and water. Everett Shock of Washington University thinks that hydrothermal vents could be where life began, the combination of heat and gases providing the means to turn inorganic sludge into self-sustaining organic compounds.[24] Shock's conclusions are based on calculations rather then experiment but weight has been given to his ideas by Gunter Wachtershauser and his colleague Claudia Huber. The two German chemists have been playing with a mix of vent chemicals. Their experiments have shown that vent reactions can create organic compounds and trigger a cascade of other reactions. Says Wachtershauser: 'Our compounds are changing and changing and changing. Life didn't begin with a soup of chemicals, which cannot do anything, which are inactive. It's change that leads to life.'[25] The theory, though persuasive, has come under attack because the chemicals that the three scientists have concentrated on are present in very small quantities around the vents and it's therefore hard to imagine them as the fountain from which life flowed.

Son of Beach

This leaves the beaches of the early Earth, places that David Deamer, a biochemist from the Santa Cruz campus of the University of California, knows well. Deamer has a lot of innovative research to his name. He has created music out of DNA using the letters used to represent the four bases as notes – A for adenine, C for cytosine, G for guanine and substituting E for thymine. He also knows what space smells like – apparently it has the same aroma as a musty old attic.[26] Primarily though he is interested in bubbles.

Deamer reasons that, although the warm shallow seas of the young Earth were a seething chemical broth, it is unlikely that chance was enough to bring the right chemicals together to create the relatively complex cellular structures we see in bacteria today. Numerous experiments have shown that naked DNA does not fare well. Alone in an environment like the sea DNA would swiftly be split into its constituent compounds and lose any ability to replicate. To be able to sustain itself and maintain its biogenic properties requires a benign, unreactive environment, in effect a cell.

Even viruses, the ultimate in hardy organisms, which are little more than a length of DNA or RNA surrounded by a protein sheath, do not prosper unless they have access to a host with a ready supply of nutrients.

With this in mind Deamer has been looking for a way in which DNA might have been protected. He has been searching for simpler cells and believes bubbles might be the answer. Bubbles enclose spaces and protect them, however briefly, from the world outside. They are places where chemicals can be isolated rather than randomly mixing together with whatever else is swilling around in the sea. They act like tiny reaction vessels and thankfully are very easy to make.

Deamer believes that the barriers or membranes surrounding every cell were once much simpler than they are now. He suspects that protocells exploited the ability of molecules known as lipids to form into bubble-like structures known as liposomes. Lipids have tiny heads and long tails and resemble very skinny tadpoles. Their heads are made up of charged sugars and phosphates, while the tails are usually uncharged carbon and hydrogen chains.

In water this charged–uncharged property proves useful. Water molecules have a weak charge of their own and the charged heads of the lipids can bind to the water molecules. Often this means that in water the lipid molecules form a double sheet. The heads join loosely to water on the outside and the tails, which are water-avoiding, poke towards the centre of the layer. When the water is dried out the sheet can buckle, close on itself and become a bubble – a precious oasis isolated from the outside world. Sometimes chemicals will be swept inside the buckling bubble, sometimes two or three chemicals will be caught up. When they do interesting things can happen. Deamer suspects the most interesting thing of all – life – happens. The membranes of modern cells are made on the orders of DNA but these membranes include a double layer of lipids.

Experiments carried out twenty years ago showed that the cycles of drying and wetting seen in tidal pools can make chemicals link into chains that are the precursors of RNA. Deamer suspects lipid sheets and bubbles were being formed and diluted at the same time – a combination that may have helped RNA, and life, get going.

Deamer was partly led to this conclusion by some work he did on a meteorite that was found in Murchison, Australia, in 1969. Deamer ground up a section of this meteorite, extracted the organic carbon within it, made it into a slurry, dried it and added water again. When he looked at the result through a microscope he found that the carbon readily formed into tiny vesicles – fluid-filled membrane sacs.

Deamer speculates that it was in such bubbles that early life got started.

Key chemical trends were well established in the early oceans. As the Isua rocks show, photosynthesis – extracting chemical energy from light – was already established. The movement of the tides and cycles of light and dark created a world-wide mixing vat in which chemicals vital to life were combined in every possible permutation. Some compounds acted as catalysts and helped the chemical systems getting established in bubbles to keep going. Once you have a self-sustaining system working inside a membrane you have a proto-cell and the beginnings of life.

All by Myself

Deamer suspects that the self-sustaining compounds likely to be lurking within the lipid bubbles were the first examples of ribonucleic acid (RNA), the precursor to DNA. Deamer has had success establishing RNA within vesicles that can keep itself going for short periods of time. Deamer and other scientists working on this problem are trying to replicate millions of years of evolution in a few years and are largely working in the dark. Unfortunately there are no extant examples of proto-cells they can draw on to see how it was done.

Around 3.5 billion years ago RNA established itself in proto-cells and the reign of the bacteria had begun. For nearly 2 billion years these simple organisms, known as prokaryotes, which lack nearly all the furniture found in cells today, had the planet to themselves. These cells have no nuclei, nor any mitochondria or chloroplasts to release energy. Despite this they are astoundingly successful.

Amoeba

It is not just chemists who are trying to find out about the origins of life. ALife researchers are just as curious about how life got going. It could be said that ALife is ahead of the biologists and chemists because in 1996 one man succeeded in getting 'life' to emerge spontaneously from a pre-biotic broth of computer code. Despite the best efforts of David Deamer and others, nothing comparable has been achieved by life scientists.[27]

The man who managed this feat is Andrew Pargellis, a research scientist at Bell Telephone Laboratories in Murray Hill, New Jersey. He devotes most of his energy to topics unconnected with ALife, liquid crystals mainly. The rest of the time he plays around with ALife. Pargellis has managed to convince Bell that ALife will one day prove useful to the company, that's why they let him tinker with Amoeba, his ALife simulator. As far as Pargellis is concerned he is just having fun.

The fun begins with a computer simulation of a primordial soup. Instead of chemicals Pargellis is using random sets of computer instructions. It's assumed that life began when the right chemicals, in the right quantities, combined to produce something unexpected. Pargellis has tried to create similar conditions inside a computer, but he is aiming to produce working, replicating programs rather than protocells. Either way you could argue that life is emerging.

Pargellis' chemical-rich sea is a chunk of a computer memory set aside for his experiment. For the sake of convenience he divided this space into 1000 separate slots and populated a few hundred of these locations with a random mix of instructions. The other slots he initially left empty.

Just like the primordial soup of the early Earth, Pargellis' random mix was rich with hints of what might be. The bits and pieces of instructions could be combined many ways. If they were combined correctly they created up to sixteen fully functioning operations that could turn useless cells into working programs. Some of the operations are dead ends and carry out no useful functions, but five of the sixteen operations are more productive. If these are executed in the right order then the organism containing them will be able to replicate itself and place a copy of itself, or daughter cell, into a vacant slot in memory. One of the defining characteristics of living things is their ability to reproduce. Pargellis wanted to find out if artificial life could arise in the same way. Specifically he was interested in: 'the creation of self-replicating organisms from an initially disordered prebiotic world.'[28]

Pargellis created only sixteen operations because this is the same order of magnitude as the twenty amino acids found in living cells. It also makes the possible number of combinations more manageable.

Once Pargellis had seeded a few hundred slots in his sea of memory with the random strings of instructions he let the computer cycle through them. In doing this it was mimicking the biochemical activity of nucleic acids that direct the functioning of cells.

If the computer found any workable instructions it would try and carry them out. Most of the instructions it came across led nowhere, or at best to a string incompletely copied to a vacant slot. In this way the mix of instructions in the broth gradually changes. As it is mixed new instructions or parts of instructions may be created, perhaps opening up a pathway to replicating, living programs.

To avoid his cyber-sea becoming clogged with dead, unreactive matter, Pargellis added a clean-up program. Every time the computer had completed 100,000 instructions this 'Reaper' program cleaned out 7 per cent of the slots and refilled them with new, randomly generated strings of code.

Using these two methods Pargellis managed to create a seething sea not unlike that of Earth's first oceans. Certainly it was a potent brew. The mutations and recombinations it encouraged have produced many types of structure. Some exist only fleetingly, others persist and come to dominate Pargellis' small world, still others are the raw materials for the fitfully alive. So far Pargellis has seen three main types of organism emerge: prebiotic, protobiotic and biotic organisms.

The prebiotic cells are incapable of reproducing, other cells that are starting to get the hang of replicating often use these chunks of instructions for their own purposes. Protobiotics manage to replicate but do so inefficiently. Often the replication process is not completed correctly and parasitic organisms in neighbouring slots commandeer this process to preserve themselves. Once the process of replication has been mastered even partially, another form or mutation is introduced. Every time a daughter cell is created in a vacant slot there is a 10 per cent chance that it will be copied imperfectly. Most of these imperfections (60 per cent) will involve the substitution of one instruction for another. More rarely a random instruction will be deleted (20 per cent) or inserted (20 per cent). Again, Pargellis put this mechanism in place to try and mimic the real world. Many chemical processes falter as they proceed for any number of reasons. Pargellis thought it only reasonable that the cells and organisms inside his world should be subject to the same capricious forces.

The final category are the true biotic, or living, cells. These replicate very efficiently. They have all the instructions needed to complete Pargellis' four-phase reproductive process. As a group they also gradually strip out any operations that perform no useful functions. These living cells are rare creations. Only in 3 per cent of cases do they appear. Often Pargellis has watched the simulation and been falsely encouraged by the direction that it is taking. A promising beginning often falters and stops long before anything that can maintain itself is produced.

There are criticisms that can be levelled at Pargellis and Amoeba. You could argue that the potential for replication is built into the program and the random mixing is rediscovering something that was there all along. Perhaps the same can be said of biological life. Billions of years ago some lucky chemicals discovered a potential to keep themselves going and here we are today. Certainly criticisms can be levelled at the relative crudity of his model. Although he uses roughly the same amount of variables that there are in cells, chemical reactions are much more diffuse and haphazard than instructions. Once a computer has an instruction it carries it out rigidly, living organisms are not so fixed. Life seems to be a much woollier process. But it has to be said that Pargellis was trying to simulate life not re-create it and abstractly he certainly seems to be on to something.

One fact certainly adds weight to his claim that he has produced a useful model. When Pargellis plotted the rise and fall of populations of replicating organisms he found a pattern of peaks and troughs similar to that found in the fossil record. Analyses of the rise and fall of species has revealed that on evolutionary timescales periods of stasis are followed by flurries of rapid change. This theory has come to be known as 'punctuated equilibrium' or 'punk eek' and it is shared by both Amoeba and the real world.

Sex and Death

Pargellis is now extending his model to let his creations have sex and evolve further. His latest attempt has separate environments for the strings of instructions to explore. Some are more hospitable than others and he hopes they will drive evolution within Amoeba much as the changing Earth provoked bacteria to diversify and evolve.[29]

Certainly sex is key to making Amoeba more realistic. Sex was discovered around three billion years ago by prokaryotes. The first sexual act was probably just the simple exchange of nucleic acids. Since then the advantages of sex caught on and soon all the prokaryotes were busily exchanging useful chemical instructions or patching up breaks in the strands of DNA and RNA they were using.

This was sex but often it did not instantly lead to the production of offspring. Instead the genes gained were used straight away. The advantage of this kind of sex is that it is much faster than the other kind where change is brought about largely by mutation rather than recombination. Innovations can spread far quicker through a population of bacteria than they can through a population of dogs.

Bacteria reproduce by budding or by growing to twice their size and then splitting. There is no direct gene transfer

involved in bacterial reproduction like there is in the reproduction of larger organisms. The next generation of bacteria
may be more robust because of the other bacteria their
parents swapped genetic information with.

Prokaryotes are much more promiscuous than larger animals who can only swap genes within rather than between
species. Prokaryotes merrily swap information all the time, a
state of affairs that has led some to suggest that these simple
unicellular creatures should not be considered as different
species. All the different types of bacteria can share genetic
information. Sometimes this information makes its way into
the organism's main DNA strand and gets passed on to its
offspring, sometimes it just hangs around inside the microbe
performing a useful task. This ability to pass around information means they are in effect one species that is, to all
intents and purposes, immortal.[30] For everything else sex is
about death. Probably the only thing that can match prokaryotes for fecundity and ease of exchange is computer software. While not all software is written in the same language
the ways in which different programs can be made to work
together is increasing rapidly. Some think that most software
will eventually be made up of short lengths of code that can
work with each other. This similarity between microbes and
software is sobering when you consider what prokaryotes
have done to the planet and how long they have been living.
It's an insight some ALife researchers are keenly pushing.

The simple prokaryotes had the world to themselves for
nearly two billion years. If you were taking that car journey
through history make sure you have great conversationalists
as driving companions. For the first two days there would
be nothing to see. The only thing you might come across
would be mats of mainly cyanobacteria, the blue-green

algae. These microbial mats trap silt and mud, with new layers growing on top of the old until gradually the banded rocks called stromatolites are formed. Today these are the oldest of life's relics.

The view does not get much more interesting even when the eukaryotes arrive. Unlike the simpler prokaryotes, these organisms possess a nucleus and several other internal structures that either store energy or help keep the cell functioning. They probably arose as symbiotic relationships developed between simpler organisms. They may have emerged first in microbial mats when one bug accidentally enclosed a fellow and found life easier as a result. Lynn Margulis, a professor at the University of Amherst in Massachusetts, has collected a powerful body of evidence to support the thesis that the earliest eukaryotes were symbiotic collections of prokaryotic cells.

Eukaryotic, or nucleated, cells are very different from their simpler ancestors. They are up to a thousand times larger and their genetic material is contained within a discrete capsule called a nucleus. The rest of the cell body encloses the energy converting and producing structures needed to keep the cell alive.

These cells first appeared around 1.4 billion years ago. They may not have added much to the view but in other respects they exerted a profound influence. By the time they do make their appearance it would be possible to roll the windows down, as the atmosphere was beginning to contain significant amounts of oxygen.

Humans may worry about the pollution we cause but this is nothing compared to the havoc wrought by the cyanobacteria that were at this time pumping out oxygen. In effect they poisoned the atmosphere for themselves.

Carbon dioxide used to make up over 95 per cent of the atmosphere but now it counts for only 0.03 per cent.[31] Gradually over the last few billion years the amount of carbon dioxide in the atmosphere has been depleted as photosynthesizing organisms have converted it into carbohydrates. Photosynthesis also involves breaking water down into hydrogen and oxygen. To many bacteria oxygen is a virulent poison and as the gas became ubiquitous it probably killed off huge numbers of them. Just as any serious environmental threat does it also spurred evolution and the appearance of eukaryotes. Many of the structures found within these cells use oxygen and appeared for the first time during the oxygen holocaust of two billion years ago. The gradual reduction in the amount of carbon dioxide in the atmosphere has driven evolution in many different ways.[32]

For the next 900 million years the prokaryotes and the eukaryotes shared the planet until the first recognizable multicellular organisms appeared. The symbiotic trend established by eukaryotes enclosing other structures is a key trend. For multicellular organisms are in reality highly symbiotic creations. All the cells in a dog contain the same genetic material but each cell type only makes use of a certain part of that instruction book. There is an unwritten agreement to work together even though cells do different jobs. In contrast prokaryotic cells have to do everything themselves.

Since the appearance of multicellular creatures life has literally exploded. The boom in multicellular animals that took place 570 million years ago is known as the Cambrian explosion. The emerging creatures had no competition for the available environmental niches they preferred, they were the first to settle into the new home. Everything was up for grabs.[33]

How exactly multicellular organisms arose is as big a mystery as the origins of life itself. While it is known that multicellular organisms share genes that dictate body plans and divisions into front, middle and back, no one is sure how these genes arose, nor how exactly they do their job.

The view from the car improves considerably once multicellular organisms appear and start fighting for space. If you were driving along a coastal region, all kinds of organisms would be splashing around in the rock pools and tidal waters, taking the air for the first time ever.

With the appearance of multicellular creatures you might be forgiven for thinking that this is the whole story of living things. Richard Dawkins' book *The Selfish Gene* implies that all you need to know about is the actions of genes and that diversity of shape and habit are merely down to genetic differences. But genetic determinism can only go so far. Much of what we can become might be contained within our genes but other forces have to be taken into account as well.

Few biologists would claim that genes are the only things that matter in an organism's development. Genes and environment work together to determine how the organism develops and grows. To explain fully how these feedback mechanisms work needs physics and complexity theory, and ALife is playing a significant part in unravelling the forces at work.[34] Together these theories are beginning to reveal that living organisms are more than just the blind workings of genes via feedback from the environment. Physical restraints on how a sheet of cells can twist and the rate at which chemical signals can propagate determine how slime mould moves and to an extent how it looks. Equally, identical twins do not have identical fingerprints. The differences in patterns on fingers arise as the foetuses grow within the womb. Again physical forces such as rate of chemical exchange will play more of a part in this process than will genes. The same forces play a part in the development of other structures within the body just as they do in finger-prints. ALife is showing that these forces play a greater part than some scientists are prepared to admit.

You don't have to go far to find further evidence for this hypothesis that genes are not everything. Organisms were once lumped together on any diagram representing the tree of life simply because they looked like each other. Now our knowledge about genetic relationships between creatures is changing the organization of this diagram. The hyrax is known to be closely related to the elephant, for example, even though it looks more like a squirrel. In the animal kingdom there are far fewer body plans than species. This is because there are only so many ways that organisms can be built and still function effectively. If genes dominated, one would expect organisms that are close genetic relations to look very similar. That they don't implies the forces shaping bodies and form act, to some extent, independently of genes.

The chemical processes surrounding the workings of genes within cells are chaotic and resist analysis, but it is possible to consider living organisms on levels other than the genetic. The story of life is all about the accretion of boundaries, from the nuclear membrane around a nucleus to the lipid layers in a cell membrane, the discrete organs within bodies, the bodies themselves, populations and planets. It is possible to study life at these levels using computer simulations or other models known as cellular automata and make meaningful predictions about what is going on. The rest of this book takes up this story.

Life Lessons

This point that life is about more than just genes cannot be overstressed. While it will clearly be a great day when we can say that we understand how the chemical-rich Archaean seas brought forth life, it will not be the whole story. An important part admittedly, but not everything. In every sense it will just be the beginning. Life is a process and to explain and understand it fully involves seeing what it does as much as where it came from. In this way we may come to see living things in a different light. Certainly there are characteristics of living systems that go beyond bodies and that also have to be understood.

Life is a self-sustaining system. Once life gets started it is tenacious and tries to build a world fit for itself. There is no pre-ordained direction to this. Living things work to keep themselves going but at the same time they are unconsciously altering the world to make it more habitable for them. We have already seen that the prokaryotic bacteria have exerted a profound effect on the Earth.

They began with the atmosphere. The early atmosphere was dominated by carbon dioxide but the gradual growth in numbers of living organisms has slowly changed its composition. Now the ratio of gases in the atmosphere is relatively stable because the organisms on Earth unconsciously collaborate to keep it that way. The composition of a planet's atmosphere is a good guide as to whether it currently harbours life in any abundance. The atmosphere is one of the routes via which the raw materials of life are circulated around a planet. The atmospheres of Mars and Venus are made up of over 90 per cent carbon dioxide and it is a safe bet that nothing lives there now. If anything did once live on those planets it did not manage to get as well established as it did here, perhaps because those planets are too close or too far away from the Sun. On Earth the explosion in the numbers of bacteria changed everything. The change in the atmosphere meant that greenhouse gases were liberated and trapped some of the infra-red light that keeps the atmosphere warm. This has meant that the temperature of the planet, bar the geologically brief ice ages, has remained largely static since the Archaean eon. We have the bugs to thank for keeping us warm and releasing the gases that let us breathe. The whole thing is wrapped up in a self-sustaining system.

Bacteria are also thought to be responsible for regulating the salinity of the oceans and ensuring they do not get so salty that life becomes impossible. They may even play a role in the formation of rain clouds. Geologist Don Anderson has speculated that they might even be responsible for tectonic activity.[35] Although calcium is found in bones and teeth and plays a part in many processes within cells it is toxic at high concentrations in its free ionic state. One of the ways to neutralize its poisonous effects is to lock it into a safe form like calcium carbonate. If bacteria pursued this course of action this would mean that a rain of calcium carbonate would slowly fall to the ocean floor and begin to accumulate. Over the billions of years that the bacteria had the oceans to themselves a significant amount of calcium carbonate could have built up. Anderson speculates that this calcium carbonate changed the chemical composition of the ocean floor near the land margins. This may have precipitated an event called the basal-eclogite phase transition, which significantly changed the properties of the rocks in the Earth's crust so much that it helped the plates to move. While much of this theory is speculation and evidence is still being sought to confirm it, perhaps it is significant that a planet devoid of life like Mars is dead to the core.[36] The lack of life may have played a part in the planet's development.

This intimate coupling of life and environment goes much further. There are many, many ways in which living things have come to control their surroundings so they are maintained at a comfortable level. Sometimes it is just luck that helps an organism survive, at other times it is a more active process. It is not just the bacteria that have learned to play tricks like this. Throughout the history of life on this planet many instances can be found of a tight connection between life and its environment.

It is obvious that plants are not just passive participants in the living world, they are not just the backcloth against which everything else moves. They too are evolving and fighting off attempts to extinguish them. Palaeontologist Robert Bakker has observed: 'Dinosaurs held the role of large land herbivores for longer than any other vertebrate group, so there must have been a rich history of adaptive attack and counterattack between plant-eater and plant.'[37]

The biggest development in plant life during the reign of the dinosaurs was the appearance of flowering plants. Bakker contends that plants started this wildly successful method of reproducing – today flowering plants are the most abundant of fauna – as a response to the depredations of the dinosaurs. Bakker believes that plant-eating dinosaurs were the fastest evolving of all the dinosaur groups and as such would have been a significant threat to the survival of plants if the plants had not evolved some way to preserve themselves. The ability to reproduce via seeds was that way.

Life has continued this trend of making the Earth more habitable for itself as it has colonized new habitats. Photosynthesis was one of the first tricks of self-maintenance that living organisms evolved. But not all plants photosynthesize at the same rate. Some do better at lower levels of carbon dioxide. As the amount of carbon dioxide in the atmosphere has dwindled this is a sensible ploy. Nearly all trees, broadleafed plants and a few grasses are known as C_3 plants and need higher levels of carbon dioxide to sustain themselves than newer C_4 plants – many of which are grasses.

This change in plants may have had a devastating effect on many species of early horses that were alive six million years ago when C_4 grasses started to make an appearance. Steven Stanley at Johns Hopkins University in Baltimore has found that C_4 grasses have nearly three times as much silica in them as the older C_3 grasses. This could have caused problems for some of the species of horses living at the time. These animals had developed long teeth that were gradually worn down as they chewed grass. But their teeth wore out much more quickly as they were forced to eat the grittier, harder-to-chew grasses that were coming to dominate the planet. It is perhaps just another way in which life and Earth are shown to be locked in an intimate, tumbling embrace.

Earth scientist James Lovelock predicts that suddenly these grasses will take over as they will be the only ones that can make a decent living out of the changing composition of the atmosphere.[38] It is a salutary reminder that the rhythms of life run deep and strong and of the intimate relationship between living things and the place they make their home.

Chapter Two

The Game of Life and
How to Play It

Now I know what it feels like to be God.
(Lines cut by censors from many prints of the 1931 film
of *Frankenstein* on the grounds of blasphemy)

The Anasazi Indians are a mystery to many anthropologists
and archaeologists. The earliest settlement built by these
Indians dates from AD 100 but no one knows where they
lived before they set up home in north-eastern Arizona. No
one knows why they picked this region, why they scattered
their villages across thousands of square kilometres of
this territory or even what they called it. What is known
about the Anasazi is that over a period of 1200 years they
established a flourishing culture in this territory, building villages, shrines and farms that have left traces of the
Anasazi's passing but few clues that help unravel their
mystery.

Some of most distinctive Anasazi settlements centre on a
region now known as Four Corners. It bears this name
because it is where the borders of Arizona, Utah, New
Mexico and Colorado meet in a neat cross. No one knows
what the Anasazi called this place nor even the name the

Indians had for themselves. 'Anasazi' is a Navajo word meaning 'ancient ones'.

Some of the Anasazi's largest pueblos can be found in and around Four Corners. Those found in the Black Mesa and Long House Valley in Colorado are among the most striking. Indeed, the valley itself takes its name from the homes built there by the Indians.

The land they chose for their home can be a hot, unforgiving country and it is bounded in most directions by desert or mountains. To the south-west are the trackless wastes where the first atomic bomb was tested. Despite the harsh conditions, the Indians prospered, gradually turning away from their crude hunter-gatherer past and embracing agriculture, masonry and pottery.

The Anasazi culture lasted over a thousand years and enjoyed a golden age lasting just over a hundred years towards the end of the thirteenth century when the Indians were at their most numerous, but, unbeknownst to them, time was running out. A period when you might be forgiven for thinking they could beat off most threats to their livelihood.

Towards the end of the thirteenth century, after an unbroken 1200-year period of occupation of the region around the Shonto Plateau, the Indians were building distinctive stepped stone dwellings, some of which were four storeys high and contained up to 1000 rooms. Beneath them were subterranean chambers used for religious ceremonies. By this time they had also mastered the arts of pottery-making and weaving. The clay pots they fashioned during this time, and the clothes and rugs they wove from cotton and yucca fibres, are decorated with striking, original designs.

Then, just as quickly as they arrived, the Indians disappeared. In the space of fifty years they abandoned the elaborate dwellings they had built into the canyon walls and the farmland where they had cultivated cotton, maize, grain and yucca for generations. Suddenly they struck out, travelling south and east to the Rio Grande and Arizona's White Mountains. On the way they seemed to lose their skill with masonry. The later settlements they established never matched the grandeur or assurance of the valley and canyon lodges. It was as if the Romans suddenly abandoned Rome and moved to Naples.

No one knows what caused the Indians to abandon the place that had been their home for over a thousand years – a place that was deeply woven into their rituals and culture. Perhaps the twenty-three-year drought that marked the end of the thirteenth century ruined the farmland, killed off their crops and made the game animals seek greener pastures. Perhaps it was the depredations of the nomadic bands of Navajo and Apache that were moving in from the north that drove them away. No one is sure.

After the exodus the numbers of Anasazi Indians dwindled to a third of what it was during their heyday, partly because of the upheaval of leaving their homes and changing their lifestyle. It was also partly because of the zeal of Spanish missionaries and troops during the fourteenth and fifteenth centuries who were looking for converts. This Spanish influence watered down the millennia-old culture of the Anasazi. Soon it became a shadow of its former self.

Descendants of the Anasazi survive to this day and are known as modern Pueblo Indians, but all that survives of their culture is their skill in pottery and agriculture. The ancient history of the Anasazi is lost.

But now the Indians are living again. The tribe has been

resurrected and Long House Valley is once again dotted with Anasazi settlements. The only difference this time around is that every member of the tribe is deaf, blind and limbless. Despite this the behaviour of the tribe is indistinguishable from its earlier incarnation. They farm the same crops, hunt the same animals, weave the same clothes and choose the same sites for settlements. Everything looks normal and every tribe member fulfils the role they have been given, be it hunting, gathering or weaving.

Watching the Indians go about their daily life in conditions similar to those of the thirteenth century demonstrates how accurate this representation is. Once again the Anasazi gradually migrate away from smaller, separate villages into larger settlements that centre on the communal canyon houses. For centuries they maintain and expand their culture. The only difference is that they do not disperse, break up and migrate. Those studying the Anasazi are no closer to an answer though. Just as no one is telling why the original Anasazi upped and left, no one can work out why this lot are staying. What makes it worse is that this time the Indians cannot talk so they can't be asked.

It should be obvious that in this recent resurrection the Indians are not Indians at all. Each community is actually a coloured dot on a grid laid over a computer re-creation of Long House Valley. Everything the dots do is determined by the rules programmed into the computer, yet eerily this makes them act just like the Indians. They live, love and leave or not . . . just as the real Anasazi did.

This simulation of the Anasazi culture has been created by an eclectic group of scientists including anthropologists, archaeologists, computer modellers and experts in complexity theory. The research group was brought together at the

Santa Fe Institute by renowned archaeologist George Gumerman. This influential, and now retired, scholar undertook the task because he was interested in simulating the living and environmental conditions of the Indians as faithfully as possible. He has drawn on the expertise of the research group to build an authentic re-creation of Long House Valley on computer. He is getting help from the likes of Jeff Dean of the University of Arizona's Laboratory of Tree-Ring Research. For years Dean has been reconstructing ancient weather patterns using data from tree rings and other climatic indicators. The researchers have drawn on this data to make their computer-generated environment match as closely as possible the conditions the real Anasazi Indians were contending with.

What the group is seeking to understand is the pressures on the Indians in order to tease out what might have made them suddenly abandon settlements that must have taken decades of sweat and toil to build. From this they hope to learn valuable lessons about the evolution of cultures.

The computer program being used by Gumerman and his colleagues to simulate the history of the Anasazi is an updated version of a much simpler, and older, program known as a cellular automaton. In fact cellular automata are only a few years younger than the electronic computer itself and they both share the same father: John von Neumann.

Brain Box

The word 'genius' is bandied about rather too much these days, and is often applied to those that really do not deserve the honorific. The people who do deserve to be known as geniuses are few and far between but John von Neumann is definitely one of those people.

Janos, later Johann and, finally, John von Neumann was born in Budapest in 1903 to a wealthy banking family. The 'von' in their name was more often used by members of the German nobility, but the title had been granted to the family by Kalman Szell, who became prime minister of Hungary in 1899. Prior to this he had been head of the bank where von Neumann senior worked.[1]

From an early age young Janos showed an obsession with finding the right way to do something. Before he was ten, according to one biography, 'he would attempt to resolve emotional issues by insisting on correct form, either proper social form or statements in logical form'.[2]

It was also during his childhood that his remarkable intellectual gifts began to show themselves. Von Neumann's parents and tutors found that he could soak up facts and solve problems at astonishing speed, all the time aided by his prodigious memory. Later in life he liked to show off his memory by recalling long lists of names from phone books, obscure facts about Byzantine culture and reams of risqué stories and jokes.

By the age of seventeen he was attending the University of Berlin to listen to Einstein lecture on statistical mechanics. Von Neumann also spent time at Gottingen talking to the most famous mathematician in the world, David Hilbert. Although years separated the two men, they spent hours talking through their ideas.

In 1926, aged twenty-three, von Neumann received a diploma in chemical engineering from the Technische Hochschule in Zurich and a PhD in mathematics from the

University of Budapest. He began lecturing at the University of Berlin the same year, the youngest man to do so.

Von Neumann carried out pioneering work in fields as diverse as economics, quantum physics and pure and applied mathematics. In every one he either solved a key problem, wrote a book that is still considered a standard treatment, or else published papers that set researchers off down avenues of inquiry that are being explored to this day.

When he was twenty-seven von Neumann was appointed as one of the first visiting lecturers to the Institute of Advanced Study at Princeton University in New Jersey. The other was Albert Einstein. Three years later he became a full-time professor at the Institute of Advanced Study, a position that he kept until his death in 1957.

Von Neumann's life was cut short by prostate cancer when he was just fifty-three years old. Although von Neumann could have done nothing with his life, he had a substantial private income, there is something tragic about the way he crammed more into his curtailed life than most of us could fit into several existences. Perhaps he sensed that he would not make old bones and worked harder to make up the shortfall. He apparently had little regard for personal safety, writing off approximately a car a year.[3] He also had a taste for parties, high living and female company.

Although later in life he enjoyed a luxurious lifestyle, von Neumann's childhood was not always so comfortable. Part of the reason that he chose to go to America was because of the way his family had been treated after the First World War. Hungary had not done well out of this conflict and had lost much of its territory. Widespread social unrest saw the installation of the Communist Bela Kun government that ruled the country for five months. One of the first things it did was take over the banks and put von Neumann's father out of a job. Although the government was soon disbanded, Hungary was not a pleasant place to live for some time as those who were thought to have collaborated with the Bela Kun government were hunted and persecuted.

Von Neumann's family survived intact but the experience made a deep impression. When the Second World War came von Neumann was eager to turn his formidable intellect to the service of the Allies. He became a consultant to the Manhattan Project and was involved in solving the differential equations that would ensure an explosive forced two lumps of uranium together in the right way to create a nuclear explosion.

The Second World War produced the first atomic bomb and it also produced the first computers. All over the world at this time men like Konrad Zuse in Germany, Alan Turing and his colleagues at Bletchley in Britain and von Neumann in America were making automatic processing machines for their governments. But it was only luck that von Neumann was involved with computers at all.

During the latter part of the war in a little known project centred on the Ballistics Research Laboratory (BRL) in Aberdeen, Maryland, a computer called the Electronic Numerical Integrator and Computer, or ENIAC, was being built. Like the Manhattan Project, ENIAC was being developed for the military. As its name implies the BRL was set up to help bombs fall in the right place. In the main it developed guides for gunners called firing tables. These were used by soldiers to ensure that the artillery they were firing was set at the right angle so the shell it was firing fell on target. The BRL sat on the same site as the Aberdeen proving ground – a firing range 96 kilometres south-west of

Philadelphia on the edge of Chesapeake Bay where new weapons were tested.

The scientists driving the development of ENIAC were John Mauchly and John Presper Eckert from the Moore School of Electrical Engineering at the University of Pennsylvania. A young engineer called Arthur Burks at the same department was drawing up the technical documentation. Pennsylvania was barely an hour's journey by train from the BRL at Aberdeen. During the early days of ENIAC John von Neumann knew nothing about it or what it was trying to accomplish. He found out about it in 1944 when work had been carried out for nearly eighteen months, and then his discovery was a happy accident.

After a routine consulting visit to the BRL in the summer of 1944 where he had been helping with projects unrelated to ENIAC, von Neumann was waiting at the Aberdeen railway station and got talking to Lieutenant Herman Goldstine, the army officer charged with getting ENIAC working. Goldstine soon revealed to the eminent mathematician what he was working on. As soon as von Neumann knew '. . . the whole atmosphere of our conversation changed from one of relaxed good humour to one more like the oral examination for the doctor's degree in mathematics,' writes Goldstine.[4]

Soon after this fateful meeting von Neumann visited the Moore School and this signalled the start of his involvement with computers and how they were built. At the first meeting with ENIAC's developers his first question was about the logical structure of the machine. It was the beginning of an obsession that von Neumann would pursue to his dying day, an attempt to show that both computers and living organisms were founded on a logical basis.

By the time von Neumann became involved with the ENIAC developers, design work on the computer had been finished and it was well on the way to being built. As construction progressed its shortcomings rapidly became apparent. The computer used 18,000 vacuum tubes, 70,000 resistors, 10,000 capacitors, 6,000 switches and 1,500 relays to crunch 20 numbers at a time, but it had to be reprogrammed by hand, by unplugging and reconnecting cables via a series of boards that resembled those used by early telephone operators. In ENIAC's memory were stored the numbers it was performing calculations on, the program to process them was the way the whole thing was wired up. For ENIAC to do a different job meant effectively building a new computer via the plug boards.

Von Neumann helped the Moore School and the BRL group overcome the deficiencies of ENIAC and draw up a design for a more powerful version that would be known as the EDVAC or Electronic Discrete Variable Automatic Computer. This is known as the first modern computer because its memory held both the program and the numbers it would be processing. There was no more rebuilding to make it carry out a different job, just input a different sequence of electronic signals. This innovation would make EDVAC much faster and far more flexible than ENIAC.

The five scientists pushed on with the EDVAC design. Eckert and Mauchly worked on building the memory, while von Neumann, Goldstine and Burks worked on the logical design – or architecture – of the machine. In spring 1945 von Neumann decided to write up the work the group had done so far and the 101-page report he authored, called *A First Draft of a Report on the EDVAC*, was published in June 1945. It detailed the work on the shared memory aspect of a computer and divided the internals of a computer into five functional elements: the memory, the control unit, the arithmetic unit and the input and output units. It was no overstatement to describe this document as 'the technological basis for the world-wide computer industry'.[5] Although only twenty-four copies of the report were originally printed, it was swiftly copied and found its way into the hands of nearly everyone who was interested in these new-fangled things called computers.

Because von Neumann's name was the only one on the report he got the credit for inventing the modern computer. The majority of computers today process data using what has come to be known as the von Neumann architecture. In fact, the report was a synthesis of everything the researchers were doing, that it had only von Neumann's name on it created a breach between von Neumann and Goldstine on one side and Eckert and Mauchly on the other. The widespread circulation of the report stymied any chance Eckert and Mauchly had of patenting the idea and making a mint, something that only increased the distance between the researchers.

One aspect of the report was all von Neumann though. For some time he had been increasingly interested in the connections between logic and biology. The report talks not so much about electronic circuits as logical elements that he likens to 'neurons', he even called the input and output parts of the design 'organs'. When he returned to Princeton after the war von Neumann would flesh out these ideas.

In the event the men behind EDVAC did not become the first to build a modern computer that holds data and program in the same store. EDVAC was not completed until 1952, problems storing the programming instructions delayed its completion. The honour of making the first modern computer went to Manchester University, where, on 21 June 1948, Freddie Williams, Tom Kilburn and Max Neuman ran the first stored program on a computer known as the 'Baby'. This 'Baby' nearly filled a room, but it got this name because it was the precursor to the Manchester Mark I. The 'Baby' was also called the 'Small Scale Experimental Machine' and its design was heavily drawn upon for the larger Mark I computer that achieved its first error free run in 1949.

The Logic of Life

For some time von Neumann had been interested in the relationship between life and logic. He believed that there was a profound connection between the two and that, at bottom, everything living things did was grounded in a series of axioms. Von Neumann thought that all that was required was the teasing out of these fundamental principles. Although he realized that living things were more complex than anything humans could fashion, he thought it reasonable to try and capture some of the essential qualities of living organisms in an artificial form.

In 1948 he presented his ideas to his peers in a series of lectures at the Hixon Symposium in Pasadena. His lectures drew on earlier work by Warren McCulloch and Walter Pitts – the men who thought up artificial neural networks. Von Neumann's lecture series was entitled 'General and Logical Theory of Automata' and was a bold attempt to create a mathematical theory that would allow biological and artificial information processing systems to be compared.

These lectures contain several profound insights that largely inspired the artificial life movement as it is today. Firstly, von Neumann saw that it was the manipulation of information that keeps an organism alive, allows it to beat back entropy, and that life is a process not a property. Mistakes in copying the information would mean that an organism would mutate and, if the change was beneficial, survive and reproduce. He reasoned that by supplying a machine with the correct instructions it should be possible to emulate living things.

Secondly, von Neumann saw that a certain level of complexity would be needed before a machine could be crafted that would be able to create other machines more complex than itself.

Thirdly, he saw that one of the characteristics dividing living organisms from machines is the fact that living things reproduce but machines must be made. If he was to build a machine that was to be lifelike it would have to be able to reproduce. Only then would he be able to bridge the gap between the biological and the mechanical.

In the lectures he laid out his ideas. He conceived of a vast lake in which floated a single robot and all the pieces needed to make other similar machines. The robot, or 'self-reproducing automaton', would swim about assembling copies of itself. He told the audience in Pasadena that the robot would possess a manipulator with which to grasp parts; a cutting device that could split two parts previously stuck together; a device to weld parts together and sensors to recognize the different bits of a robot. Information and instructions would be supplied to the robot in the form of girders bolted together in long sawtooth-shaped lines. Any mistakes made when the 'girder tape' was copied for any offspring might improve the functioning of the robot and provide a route for beneficial mutations.

Von Neumann never meant for this fantastic robot to be built. It was simply a thought experiment. All he wanted to do was show that theoretically it could be done and to put the whole project on a sound logical basis. But von Neumann's friend and colleague Stanislaw Ulam saw a way for the reproducing automaton to live, after a fashion. Ulam proposed that von Neumann employ a much more abstract method to realize his creation. Over twenty years before von Neumann delivered his lecture Ulam had been speculating about just such living automata with friends in the cafés of his native Lwów in Poland.

Ulam suggested that von Neumann use an array of square cells – a giant draughts board – that was programmed using rules that would make the system appear alive. Time in this abstract universe would proceed by discrete steps. At each time step every cell would be in one of many states. Its state would be determined by how many of its eight neighbouring cells were filled or empty. If the right instructions were given to the cells, a collection of them could feasibly be called an organism. The trick was finding the right instructions so that the collection of cells would evolve, adapt and flourish.

The idea had instant appeal for von Neumann. Using this approach would mean that there was no messing about with mechanical robots and supplies of raw materials. Instead he could play with the idea of an organism in the abstract and would be able to show the truth of his ideas far more easily. He could make clear that the essence of life is information.

Today von Neumann's view has gained wide acceptance. Now it is thought that the key to self-reproduction is possession of a description of oneself in one form or another. That information is then used in reproduction. Ideas of vitalism have been dropped. There is no need, or room, for ghosts in these machines. As William Poundstone, an authority on cellular automata, puts it: 'Biology is now seen as a derivative science whose laws can (in principle) be derived from more fundamental laws of chemistry and physics.'[6]

Von Neumann set about working out the rules for his reproducing automaton and the states its cells would have to go through if it was to breed and evolve. It was to be his last great work and he would not finish it before he died.

A blueprint for the cellular automaton was unveiled during a series of lectures delivered between 2 and 5 March 1953 at Princeton. In total the creature occupied over 200,000 squares. The central structure of the automaton was a box 80 by 400 squares in dimension. It also possessed a constructing arm and a tail built of around 150,000 squares. Each square within the organism could be in any one of 29 separate states. Twenty-eight of the states represented the different roles a cell could take on, the twenty-ninth state was for empty cells. To reproduce, the machine scanned the information in its tail and carried out the instructions it found there, slowly using that information to build a replica of itself in an empty portion of the artificial universe.

The creation was unwieldy and inelegant and von Neumann had no way of ensuring that it worked. He had only his faith in his own mental faculties with which to reassure others that this creation would do what he said it would. In fact, he never even got this far. Working out the theory took longer than he expected and on 8 February 1957 he died, before the work was done. The central ideas had been exposed to the public in the pages of *Scientific American* in 1955 but the manuscript that would have laid out the whole idea was not completed.

In fact, even if von Neumann had lived it may have been ten or twenty years before a computer emerged that was powerful enough to run his cellular automaton.

In 1995 a graduate student at Princeton, New Jersey, called Umberto Pesavento was studying von Neumann's automaton and was having trouble envisaging how the thing worked. To cure his curiosity and aid his understanding he updated the automaton to run on modern computers. With a little tweaking Pesavento got the program representing von Neumann's automata to reproduce. Only later did he learn that he was the first person in the world to try doing this.

While the cellular automaton runs easily on most of the computers found in universities these days, the same was not true of those around in the 1950s. Pesavento estimates that von Neumann would have been nearly 80 before computers became powerful enough to run the automaton he dreamt up. On the plus side, bar a few minor errors von Neumann got most things right. Pesavento does say that von Neumann overestimated by 10,000 times the room the automaton would need in order to reproduce. Now anyone with a reasonably powerful home computer can download the software and watch it crank through its cumbersome reproductive cycle.[7]

Von Neumann's notes on cellular automata were collated, edited and completed by Arthur Burks, one of the ENIAC developers. Burks later set up the Logic of Computers Group at Michigan, one of the first places in the world that cellular automata were run on computers and a training ground for many ALife alumni.

The collected papers reveal that von Neumann was not simply out to create any replicating system. Characteristically he thought that was too trivial and straightforward a task.[8] He was interested in creating an automaton that could represent any and every replicating automaton. He also wanted it to be capable of carrying out any computational task that any computer could do. In this he was following the work of Alan Turing (see Chapter Three). He demanded that his automaton be capable of these things because he believed this is exactly what living things do. Organisms differ in many ways. Their bodies are arranged differently, they eat different things and they live in varying habitats, yet the way genes work, the rules on which they work, are ubiquitous. Different interpretations and extensions of the same rules have led to the diversity of forms we find in the world today. Von Neumann wanted the same to be true of his replicating machine. It may use software to reproduce itself but on a fundamental level it was involved in the same kind of project. It was alive for the same reasons that you and I are alive.

Good God

Von Neumann's pioneering work on an abstract lifeform, albeit a clumsy one, is where the whole field of ALife proper begins. Yet after von Neumann died few people took up what he had left unfinished. This was partly because if von Neumann had found something hard to do, few others were going to be able to finish it. Also it was partly because at the time it was impossible to see the thing working. As Pesavento says, there were simply no computers around at the time that it would run on.

Research into artificial life and cellular automata (or CAs) languished until 1970, when John Horton Conway, a Cambridge mathematician at Gonville and Caius College, took it up. By the time Conway turned to CAs he had already established his academic reputation by making key discoveries in number theory. The research appointment this early work won him left him with a lot of time to pursue whatever subject interested him and he found CAs fascinating.

Conway did not simply take up where von Neumann left off. Like many others before him Conway saw the cellular automaton that was described in the 1953 Princeton lectures as hopelessly overcomplicated. Conway's attempt would be less grand, less sweeping, but much more useful and accessible.

What captivated Conway was what CAs might be able to do if they were easier to work with. He wondered what would happen if some of von Neumann's more exacting conditions were dropped but the creativity was preserved. As a child Conway had been captivated by games such as noughts and crosses and orange boxes, and he was interested to find out if a game could be made of cellular automata.

So Conway set about looking for a more straightforward way of doing what von Neumann had done: create a self-replicating machine. After two years of experimentation and many unsuccessful attempts, including one known as actresses and bishops, Conway came up with what is now known as the Game of Life. What took so long to develop was the set of rules that gave the game the flexibility to produce a variety of interesting forms but also allowed it to support universal computation.

In many respects the Game of Life is similar to von Neumann's original. It is played on an infinite square grid, time moves forward in discrete steps and each square is known as a cell. In stark contrast to von Neumann's twenty-nine states, these cells exist in only one of two states: alive or dead. Live cells are occupied, dead ones are empty.

The rules of this abstract universe are deceptively simple. If an empty or dead square has three of its eight adjoining cells filled, it becomes alive. If a square has only two neighbouring cells occupied, it stays unchanged. If a living cell has less than two neighbours or more than three it dies.

The Game of Life is unlike most games because players play little part in determining its outcome. Once the player has set up the universe by scattering a few living cells around, he plays no further part. He merely watches to see what he has wrought. Because it involves merely setting up a toy universe and letting it run off by itself, Conway's invention has been called the 'God game'.

The details of the game were revealed to the world via the pages of the magazine *Scientific American*, largely through the column of one of its regular contributors, Martin Gardner. In October 1970 and February 1971 Gardner wrote about Life. Part of the reason that Conway was happy to have the Game of Life written about in the press was because he was having problems proving that it was capable of supporting universal computation. Through Gardner's column he issued a challenge and offered $50 to anyone who could do this. He quickly lost his $50.

While the game was being developed at Cambridge,

Conway and his colleagues played it using counters and sheets of graph paper. While it is perfectly possible to play the game like this, it is by no means the best way. Checking the fate of each living square is laborious and makes it difficult to get a feel for what is developing and the complex forms and interactions that are taking place on the grid. Thankfully the simplicity of the game means that it was easy to program a computer to do the checking and updating. Conway himself used to play the game on a Digital PDP-7 computer.

As soon as people could play Life on computer, interest in it took off. A lot of computer scientists, students and hackers have spent a lot of time just watching the Game of Life play itself. In 1974 an article in *Time* magazine grumbled that 'millions of dollars in valuable computer time may have already been wasted by the game's growing horde of fanatics.'[9]

Games, Trains and Pentominos

What makes the Game of Life so compelling is its unpredictability. The simplest patterns of cells can give rise to the most complex forms. At times a computer screen running Life seems to resemble a primordial pond as stable organisms grow out of the dark, thrash briefly around the screen and disappear again into the murk. As Game of Life chronicler William Poundstone has observed: 'The Life universe is one of the most vivid examples anyone has found of a complex world that derives from simple premises.'[10]

The Game of Life is not really a game. Most games played on boards are won or lost because of the luck or skill of the players, but with the Game of Life there are no players, no winners or losers, everyone is a spectator to the strangest of sports. It is a window on to another mathematical universe with you in the creator's seat – perhaps this is another secret of its appeal. The universe of the Game of Life still remains largely unexplored. It is a world still awaiting its Magellans, Drakes and Darwins. ALife researchers are happy to be the ones charting these seas and cataloguing the flora and fauna they find.

A natural history of the Game of Life universe already exists. *Lifeline* – the journal compiled by fans of the game – became a record of the lifeforms that were found by those exploring the Life universe. All the objects and organisms found so far have been given names and, just as in the natural world, they are divided into species, genera and phyla.

Some lifeforms never change and as a result are known as 'still lifes'. They get names like 'block', 'loaf' or 'beehive'. Others are known as blinkers or oscillators because, though stable, persistent forms, they switch between a variety of states. Some of these go by the name of 'traffic lights' because they pass through several stages and on a fast computer appear to blink like twinkling lights.

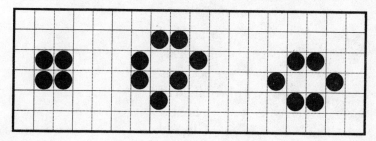

A block, a loaf and a beehive

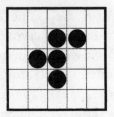

The 'r' pentomino

Still other shapes are mobile and slowly move around the Life grid. One five-cell shape called a 'glider' passes through a four-stage process that moves the whole shape one square diagonally. Also discovered in the Life universe are forms that go by the name of 'starships', 'glider guns' and 'puffer trains'. Some of these objects are 'natural' because they often arise naturally in the Life universe. Others are 'engineered' because they have never been seen to occur naturally and have to be constructed for their properties to be investigated.

One astonishingly fecund shape is the 'r' pentomino. It goes by this name because it is a five-cell edge-connected form that looks a little like a lower case r (see above). Despite its unassuming appearance the 'r' pentomino is capable of producing an astonishing number of patterns. It takes 1300 generations for the 'r' pentomino to settle into a stable state during which time it will have produced:

8 blocks
6 gliders
4 beehives
4 blinkers
1 boat
1 loaf
1 ship

So far, so whimsical, but this does not seem to have much to do with establishing a link between an artificial universe and the real world, or between living things and squares on a computer screen.

Von Neumann's cellular automaton was so complex partly because he was patterning it after the computers he had just finished working on and partly because he wanted it to adhere to specific mathematical proofs. These established that it could emulate any self-replicating machine and that it could act like a universal Turing machine. This theoretical machine was developed by the British mathematician Alan Turing and was capable of emulating the processing capabilities of any other machine.

Conway thought that the most important aspect of von Neumann's automaton was the fact that it could act as a universal computing machine. This is part of the reason that it took a long time to find the ideal set of rules for Life. Conway had to tinker in order to ensure his version did the same job. In doing this he was trying to establish a profound connection between living things and the logical world.

By picking the right rules Conway believed that he was, in some less complicated way, re-running a process that takes place in our own universe. At some point (perhaps Planck time) everything was set up, the rules were established and the whole thing was left to run. The same rules determined what happened to new generations of organisms in our universe. The use of the same rules over and over again, recursively, has once again produced ships, loafs and beehives. This time though they are not made up of squares on a computer screen; instead they are tea clippers, French sticks and homes for honey-making insects.

From Universal to Universe

Cellular automata are at the heart of ALife. The ideas underpinning them are what makes ALife worth doing. They forge a link between the complexity of the world we find ourselves in and an artificial world that is easier to study but reveals much about both.

The central point is this: living organisms are physical systems made up of elements operating to a set of rules. There is no need, or room, for non-physical forces such as souls to get involved. Some of the rules operating in living things we know about, others are proving harder to grasp and uncover. The number of rules operating, the number of elements involved and the ways they can interact make it difficult to understand what is happening in actual organisms. The blooming, buzzing confusion makes it hard to pick out what is relevant and what is superfluous. What we need is a way to remove all the extraneous elements to reveal the essence, what it is about living things that keeps them going. CAs are that way. They remove the muscle, flesh and bone and leave the essential rules exposed. Although in the case of CAs the number of rules operating are fewer than in living organisms, nothing, apart from the superfluous, should be lost. Despite the fact that there are only a few rules the complex forms that result and the fact that they emerge out of a system that is computationally universal lends weight to the claim that by studying a CA you are studying life: life in the raw and life that is much more tractable than doing experiments with fruit flies, cats or rats.

The connection between CAs and life is not purely computational. There is evidence that some cells in the body act like computers and process information.[11] A convincing case can be made for a link in this sense but it is not one that CA researchers pursue. They leave that to the robot makers (see Chapter Three). The connection is phenomenological, which essentially means that living things and CAs do the same things. The complex dynamic activity you find in fruit flies, societies and cells is also found in CAs. Using CAs to model these moving targets makes them easier to study.

Boston University physics professor Tom Toffoli thinks that the only reason CAs are worth studying is because of this connection with real life. Toffoli says that mathematically CAs are not that interesting. Although it must be said that the mathematics of CAs are easier to work with than those of other dynamic systems like water flowing round a propeller or air over a wing.

Toffoli is another alumnus of Arthur Burks' Logic of Computer group at Michigan. It was there that he did his PhD thesis on what CAs were worth saving for.[12] The conclusion of his thesis was that the saving grace of CAs was their close affinity with reality. Before Toffoli did his research many thought that CAs had little connection with the real world because the basic physics of the two worlds differed widely. But Toffoli was able to show that in fact there were strong affinities between the two. The connection showed that both worked in the same way. As such they could legitimately be used to model and study problems in this field.

Once he had got his PhD Toffoli became the first staff member of the newly formed Information Mechanics Group at MIT. The research group was set up by Ed Fredkin – one of computer science's oldest hackers who has made enough money out of the field to buy his own Caribbean island. Fredkin also has an abiding interest in CAs. He set up the Information Mechanics Group as a way of exploring his contention that all physical phenomena are rooted in information.

At MIT Toffoli hooked up with Norman Margolus and got to work on developing a dedicated CA computer. They needed a faster way to crank through the time steps of a CA. Even a powerful general-purpose workstation can only crank through a million-cell universe every minute or so;[13] far too slow to do anything useful. This was especially true for the researchers because one of the problems they were interested in tackling was turbulence in fluids, which has long been mathematically intractable. Fluids are made up of many trillions of interacting particles so to be able to model these in real time demands a computer that can work on the same scale.

A decade of work has produced CAM-8, an eighth-generation cellular automata machine that can update a CA quicker than a supercomputer could. CAM-8 deals with a million million cells at a time and is being used to simulate crystallization, reaction–diffusion systems and fluid flows. Modelling these types of physical systems on conventional computers or even supercomputers is usually an exercise in frustration. The numbers of particles in a liquid and the number of ways they can interact are staggering and eat up computer time. CAM machines mean anyone can study such things easily. The particles in fluids can be represented in such detail that Toffoli now considers that he is not working with a computational system but instead he says he is manipulating 'programmable matter'. In his opinion he has demonstrated that CAs possess some of the same properties of the stuff found in the real world but with the advantage of being easy to manipulate and set up to tackle a particular problem.

Class Action

The link with the real world may not stop with fluid flows and studies of turbulence. There are those who think that many biological phenomena, such as the markings on some shells, the growth of snowflakes and even the onset of cancer may be evidence of CAs at work.

Stephen Wolfram is the man looking for the influence of CAs in biology. British-born Wolfram is an intellectual wunderkind who went up to Oxford University aged 16 and got his doctorate from Cal Tech aged 20. At Cal Tech he kept very exclusive company, working with Nobel laureate Richard Feynman and theoretical physicist Murray Gellmann. After leaving Cal Tech Wolfram took up a post at the Institute of Advanced Study,[14] where von Neumann worked when he first thought up the idea of CAs and a fitting place to continue the investigation into their properties.

Wolfram declares himself frustrated by the slow progress of biology to reveal how living processes are organized and the way they work. He has strong opinions and is contemptuous of most of biology, seeing it as mere naturalism rather than an investigation into fundamentals. 'What we call life is something that is defined more by its history and heritage than its properties,' he says. The problem with life is that it is hard to know what is important and what is not. Wolfram says that while models of living systems can be built, typically they have not been very successful, usually because they are over-complicated. Wolfram thinks that CAs may be a way to investigate those properties and gain a much better understanding of how life is organized. Wolfram says that by analysing CAs 'one may, on the one hand, develop specific models for particular systems, and, on the other hand, hope to abstract general principles applicable to a wide variety of complex systems.'[15]

Before moving on to fundamentals though Wolfram had to do a little naturalism of his own. He had to classify the different sorts of patterns CAs can produce.

CAs can differ in several ways. They can be two-dimensional like the Game of Life and have their rules operate both horizontally and vertically. In such a CA the eight squares surrounding each cell determine the fate of the central cell in the next time step. A one-dimensional CA works only horizontally, with each generation plotted on a new line. Only the cells to the right and left of an individual cell affect its fate in each generation. The history of a one-dimensional CA can be grasped in a glance because every generation is preserved on its own line.

CAs can also differ in their rule sets. John Conway experimented with a lot of different rules before he found the perfect balance between expansion and order. What changes is the fate of the central cell given the state that its neighbours are in. In some rule sets two live neighbours means a living cell dies, in others it will mean that it stays living. Subtle changes in the rules can have profound effects on the patterns produced by the CA.

The first six generations of a 1–D CA

The one-dimensional CA on which Wolfram did his initial work had 256 different rule sets. Wolfram seeded the first line of his CA with a random number of live cells, then tried out the different rule sets to discover the kinds of patterns they produced.

He found that CAs can be divided into four different classes. The first sort of rule sets produced patterns that quickly died out. The second class found a stable form and reproduced it for ever. A third type produced chaotic patterns that keep growing. The patterns produced are fractals. Patterns that look the same at different scales. The fourth sort produced patterns that never settle down and grow and contract irregularly. The classes that Wolfram identified do not just apply to CAs, they also seem to have parallels with the behaviours seen in dynamic systems.

Class I CAs swiftly reach what are known as limit points, effectively dead ends. Class II CAs produce patterns that are very similar to the small, discrete cycles like gyres, eddies and waterspouts that sometimes appear in liquids and gases. Class III CAs tend to produce patterns that are chaotic and never settle down; global weather patterns are the classic example of this kind of dynamic system. Class IV CAs are hard to define. They are easy to identify when set against the other classes of CAs but as Chris Langton, another long-time CA fan, says: 'no direct analog for them has been identified among continuous dynamic systems'.[16] They produce complex structures that persist over long periods of time, like life.

Wolfram's survey was empirical rather than analytical but it seems to have revealed some qualitative differences between CAs. His results have been independently corroborated and extended by others in the ALife field. The survey has important implications for anyone who believes that CAs can be used to study complexity or analyse some biological phenomena. Wolfram believes that organisms often unknowingly employ CA-type systems to create some of their characteristics. One of the most striking examples of this can be found on the shell of the mollusc *Natica enzona*. Anyone comparing the patterns on the shells of these molluscs and some of the patterns that form in a CA would be hard pressed to deny some sort of connection.

If it is the case that recursive systems are widely used in nature, then starting to classify them could also help us to understand them. At the very least they could be used to model them. It is unlikely, though, that living systems use only one sort of dynamic system. Wolfram is convinced that Class III and IV CAs are most lifelike.

Fractal patterns are found throughout the natural world. In 1997 physicist Geoffrey West and ecologists James Brown and Brian Enquist published a paper in *Science* that explained, as they put it, the 'template of life'. One of the great puzzles of living organisms is why all animals obey the same simple law for metabolic rate. Whales may be millions of times larger than mice but they do not consume millions of times more energy. The energy demands of an animal follow a consistent and predictable formula called Kleiber's law, but until 1997 no one knew why. West, Brown and Enquist found that fractals were involved. The trio built a model that sought to explain why this was so. They reasoned that many multicellular living things are sustained by the transport of food and/or energy through a branching transport mechanism such as the circulatory system. It turns out that the most efficient way to move these materials around is through a fractal system that splits and grows exactly like the blood vessels in the bodies of animals. Some animals may have larger versions but all the examples of lungs or blood vessels or other transport systems are different only in terms of scale. Some plants use a similar system to move nutrients around. This explains why the branch patterns of some trees bear a remarkable similarity to the pattern of airways within the lungs.

There is a limit to how small the finest gradations of this system can go that is dictated by the size of the particles passing through them. With blood vessels the size of red blood cells is the limiting factor but even then these cells can be squashed slightly to enable them to slip through narrow capillaries. Even so, as the system grows and is used in larger animals it remains the same at different scales. Given this it is no surprise that the energy needs of animals grow predictably as the size of the beast increases.[17]

Distinctive fractal patterns have also been found in breast cancer cells that should make it easier to spot malignancies. When a cell becomes cancerous the DNA in its nucleus forms into irregular clumps. Oncologists familiar with cancer cells can often spot this clumpiness but sometimes they miss the tell-tale signs. Now researchers have noticed that the clumpiness of the DNA is fractal at a wide range of magnifications and are designing software to spot this characteristic.[18] They suspect that it might prove useful in diagnosing other types of cancer as well. Fractals are also found in the patterns of chemical signals that neurons transmit.[19] Clearly fractals are widespread throughout nature. Perhaps what we are seeing in these examples is the playing out of a huge CA-type system. If so ALife will have made a startling contribution to the study of living things and could help us unpick those highly complex biological systems and home in on the important variables.

Wolfram for one is convinced that the central point about CAs will prove its worth even if the truth of it has so far failed to penetrate the scientific community to any great degree. 'It is an intuition that has not been fully absorbed,' he says. 'The main point is that you can have a very simple rule and initial conditions and from that can effortlessly build behaviour of great complexity.'

Where Life Thrives

CAs may also be able to help us appreciate some other facts of life such as the conditions under which it flourishes, what drives evolution and the best places to look for it.

The man who blazed this trail is Chris Langton, a pivotal figure in the ALife movement. It was Langton who coined the term Artificial Life, organized the first ALife conference in 1987 and in doing so single-handedly brought together a group of people who did not know they had common research interests until Langton showed them they did. Without him there would be no ALife movement.

Langton carried out a survey of CAs very similar to Wolfram's although he based his conclusions on thousands of runs rather than just hundreds. He did this because: 'Classification is not enough. What is needed is a deeper understanding of the structure of the observed classes and of their relationships to one another.'[20]

Developing this deeper understanding led Langton to discover a key metric, which he dubbed lambda, that is correlated to the ease with which information can flow in the artificial world of the CA. If, as many people think, life depends for its survival on processing information, then lambda could be a very important measure. At the same time he has revealed how life manages to stay alive and beat off the attempts of the Universe at large to dismantle it.

Consider a 2-D CA, such as Conway's Game of Life. In that CA the fate of every cell in the next time step is determined by two factors: the state that cell is currently in, i.e. on or off, alive or dead; the state of the eight neighbouring cells around it. Each of the nine cells we are considering can be in any one of two states, so at any time the neighbourhood will be in any one of 2^9 possible configurations. We can completely define everything that can happen in this corner of our pocket universe by constructing a table with 512 entries. Each entry in the table represents one configuration. This effectively defines the physics of this neighbourhood.

The configuration of the neighbourhood determines the fate of the cell at its heart. In some situations it could give rise to a living cell and in others a dead cell. In our example CA there are 512 ways of filling in this table. Lambda is the ratio of how many living cells or '1s' there are in a given CA divided by the total possible number of results, in this example 512.

Langton's exploration yielded several interesting results. Firstly he found that Wolfram's four classes were in the wrong order. As the value of lambda increases CAs pass through several distinct states, just as Wolfram observed. But the order goes I, II, IV, III. The CAs that produce the complex patterns come before rather than after ones that give rise to chaotic patterns. The difference between the two is subtle. 'Complex' is a difficult concept to define. Intuitively we can grasp that people are more complex than flies but it is hard to formally define how. Flies and humans have a lot in common, DNA for a start, but their behaviours are very different. We can do some things that flies cannot, like algebra, but there are many things, such as flying through bushes, that humans cannot match. It's hard to say which one is more complex. 'Chaotic' though is easier to pin down. There are measurable characteristics that reveal how chaotic a system is. Some of these measures are mathematical, one is called a Lyapunov exponent and can reveal a system's sensitivity to initial conditions. Exquisite sensitivity is one of the signal states of chaos. Other characteristics reveal themselves when chaotic systems are modelled on computer. When modelled in this way patterns can be found that are invisible in the real world system. Anyone simply observing something like the weather might be forgiven for thinking that it was largely random and it was only possible to predict what would happen next on a very short time scale. In fact, order is hidden within the weather and it is anything but random. The most interesting systems on Earth, the living ones, are a tricky mixture of both complexity and chaos.

Wolfram's classification was purely empirical so this reordering was not overturning any great insight or long-held belief. Shuffling the coins in your pocket does not change their value, merely the order in which you pick them out. Langton thought that the fact that the classification system should be ordered differently after a more systematic search was highly significant.

The changing lambda value gave rise to patterns that fit very well with Wolfram's classification system. CAs that have a low lambda value are not very interesting. If any patterns emerge at all they quickly die out. As lambda increases the universe of the CA becomes busier and patterns begin to propagate. The most interesting CAs, those that correspond to Wolfram's classes III and IV, have lambda values that centre on 0.5, when a balance between 0s and 1s, living and dead, is emerging. In these CAs patterns, rhythms and stable structures emerge. Then as lambda approaches 1 the CA becomes increasingly unstable and any structures that do emerge are quickly consumed by the frenetic pace of life and death.

At lambda values around 0.5 the behaviour of the CA is either complex or chaotic. But Langton realized that the most interesting parts of all are at the regions between these recognizable states. As the CA passed from one region to another descriptions of what was happening got longer, once within the new region fewer words were needed to describe what was happening. Only in this transition region are the stable and propagating structures like gliders seen. It is the place where universal computation can take place. It is the crucible of life.

Langton called this sweet spot a 'critical phase transition' and it is not just a phenomenon of the CA universe, similar events are found in the natural world too. When substances are changing from solid to gas they pass through just such a region – their liquid state. Liquids are very difficult to study because they exhibit such complicated behaviours. What is important about liquids is not what they are made of but the way that the elements within them start to organize themselves. At the point of a 'phase transition' the chemical properties of the particles start becoming largely irrelevant. Instead the information each one carries becomes more important, local interactions no longer determine how the fluid acts. Interactions take place across the fluid, the movements of the whole mass become coordinated and organized. How exactly this happens is still being investigated but evidence for it is overwhelming.

Exactly the same change is taking place in a CA at the point of a 'phase transition'. The actions of the elements are starting to become coordinated across the whole structure. This has led ALife researchers to speculate that CAs are probably very good models of living organisms. This affinity is bolstered by the fact that CAs deal only with information. There are no chemicals present in a CA and no thermodynamics but these experiments show that this does not matter. If the essence of an organism is the information within it then something that deals with nothing but information is probably a good model. All that is important at this phase transition point is organization, CAs are abstracted away from chemistry, but they lose nothing in the process.

It is not only ALife researchers that are interested in phase transitions. Work on phase transitions has revealed that at critical junctures in the development of a system organization spontaneously emerges. This behaviour has been found in a huge variety of systems. Everything from heart attacks to earthquakes, brains and businesses have been found to be on or passing through phase transitions at some point in their development. Some systems, usually living organisms, seem to be stuck in this state. Because of the ubiquity of this phenomenon the field of research it is part of is called 'universality'. It implies that at certain times many different systems act in exactly the same way. The individual elements within them cease to matter as information dynamics takes over. It means that studying life with CAs, as long as they are of the same universality class, is a valid enterprise because in some situations exactly the same thing is happening in living systems as is happening in a CA.

Research into the dynamics of hydrogen-bond networks found in water has revealed very complicated structures and interactions. Langton speculates that these might have acted as a template for the organic compounds swilling around in the Hadean seas and given life something of a jump start.[21]

The implications of this insight are far reaching. It removes the distinctions between living organisms and how they organize themselves and computer hardware and software and how that is organized. As Langton says: 'However, if it is properly understood that hardness, wetness, or gaseousness are properties of the organization of matter, rather than properties of the matter itself, then it is only a matter of organization to turn "hardware" into "wetware" and, ultimately, for "hardware" to achieve everything that has been achieved by wetware, and more.'[22] A two-way street has been established. There is no barrier to computers starting to take on the properties of living things and living things often have the properties of computers.

What this also gives clues to is how exactly life got going in the first place. Vitalism was the philosophy that there was some lifeforce that kept living things alive. While no one believes this any more it has proved hard to say just exactly what should take its place. Information is looking like a good bet.

Langton says that there is a lot of evidence that life uses information to maintain itself in the critical phase transition region.

Cell membranes hover in a quasi liquid/solid state and DNA is constantly being zipped and unzipped. The brain too has to be kept within a very narrow temperature band to function normally. There is a narrow region in which life thrives. Life avoids lambda values at the extremes because at these places there is either too much or not enough information around. Equally a chaotic universe is one that life finds inhospitable. As he puts it life has 'learned to steer a delicate course between too much order and too much chaos.'[23] It is the threat that keeps us going, the threat of a plummet into chaos or being frozen in stasis that keeps us so very much alive.

Holding Pattern

Like many of the people who showed up at that first ALife conference, Langton had drifted for a while before finding out what it was that he wanted to study and a place he could study it. In the late 1960s and early 1970s he dropped out of the first college he attended partly because he was following no set subject and partly because his position as a conscientious objector to the Vietnam War put him at odds with the college authorities.

Instead of joining the army he did his time at Boston Massachusetts General Hospital, where he was put to work as a morgue porter. In his first week he received a sharp lesson in the tenacity of life when one of the bodies he was wheeling to the morgue suddenly sat up. He requested a transfer and was moved to the hospital's Stanley Cobb Laboratory for Psychiatric Research. There he was given the job of looking after the DEC minicomputer. It proved a fortuitous move.

Over the next couple of years Langton spent time doing data analysis work for the lab and tinkering with the computer in his spare time. He learnt a lot about programming and as much about the malleability of software. His interest in computers only increased when a colleague brought in a copy of Conway's Game of Life that ran on the DEC computer.

Langton spent a long time playing with the Game of Life, watching it spin through its generations and playing God by setting up particular structures and seeing what emerged from the collisions. He placed glider guns on the grid at various angles to make their streams of gliders cross and crash. Langton did not know it but he was doing exactly what the hackers at MIT had done when they tried to win Conway's $50. It was in just such a way that R William Gosper and his co-workers at MIT discovered the Game of Life was capable of producing an ever-growing pattern[24] – one of the key requirements for making it support universal computation.

The lab at the hospital wound down in 1972 and Langton moved on to more programming work at the Caribbean Primate Research Centre in Puerto Rico. In 1975 he decided to go back to school to fill in some of the gaps in his education. He took some courses at Boston University and then, on his way to Tucson and the University of Arizona to start another course, he received another sharp lesson in life and mortality. He had a hang-gliding accident that nearly killed him.

He broke his arms and legs and most of the bones in his face. He was hospitalized for months but he used the time to read up on all the things he had missed. By the time he was well enough to complete his trip to Arizona he knew what he wanted to study: artificial biology. At Arizona he took a huge range of courses, picking those that seemed relevant to his rather esoteric interests.

He got married, bought a computer and discovered the works of John von Neumann that had been edited by Arthur Burks. Langton contacted Burks, who told him that Ted Codd had done some work to simplify von Neumann's twenty-nine states down to just eight.

Langton got hold of Codd's work and began playing with the CA. But he realized that it was still too complicated. Although Codd had simplified the system, his characterization retained von Neumann's condition of construction universality – it still had to be capable of building any other automaton.[25] Langton thought that this was an unnecessary and restrictive condition. To him what was important was computational universality, largely because this was a defining property of living things setting them apart from the stuff they were made from.

Langton tinkered with his version of Codd's CA for a long time before he got the conditions right. But he managed to simplify the system and though it still had eight states it used far fewer rules than Codd's CA. The Game of Life has two states, alive and dead, Langton's CA had seven live states and one dead state.

The final pattern resembled the letter Q, a loop with a

tail poking east. Information flowed around the inside of the Q and out towards the tail. There the instructions build the beginnings of another loop. The outside of the loop were cells in a 'sheath' state that kept information flowing along them and constrained the direction the Q could grow. Although Langton only programmed in the rules to make loops reproduce, he found that when they were left to fill a screen order emerged unbidden. Like coral, the loops established a colony with all the reproductive activity taking place on the margins. As space became filled and there was no more space for daughter loops, enclosed loops died off.

The work is significant because it rekindled interest in CAs when they were starting to lose favour and because it showed the minimum conditions for self-reproduction. Since then an even simpler reproducing CA has been created using only eight states and thirty-one rules.[26] Langton's loops are also significant because they are what convinced him he was on the right track and trying to create an artificial biology was a legitimate research plan.

Brains Without Frontiers

The work of both Langton and Toffoli has been combined in one of the most ambitious ALife projects ever undertaken, the construction of an artificial brain. The man attempting the task is Hugo de Garis, a native Australian who began his scientific career working on evolving neural networks to control robots. Now he has moved from working with tens of neural elements to millions.

De Garis is not a man given to self-doubt. He consistently talks up the prospects for his project, despite the scepticism of many in the ALife community. He does his professional standing no good during interviews when he often speculates about what the world will be like when building brains is commonplace. He was interviewed in *Wired* magazine in December 1997 and the article introduces him by saying 'Hugo de Garis dwells on the fringe'.[27] In the article he talks about 'artilects', artificial intellects with intellectual capacities many orders of magnitude above human levels. They could be as large as asteroids or the Moon. This is what he wants to build and he goes on to say that: 'The issues of massive intelligence will dominate global politics in the next century.'[28] For the moment though he is focusing on a more modest aim – creating a brain for a robot kitten.

De Garis has taken the same starting place as Chris Langton by basing his research on the work done by Ted Codd, but rather than adapt this work to make self-replicating loops, de Garis is using them as neural networks. Just like Langton's loops each element starts off isolated and uses a succession of states to extend and grow. These new structures do not separate from their parent but remain attached and form a body of nerve cells. Once they have grown for a while they may encounter the bodies of other loops. When they do they may form synapses or connections so information can flow between the two cells. In this way de Garis hopes to build up the neural pathways for an artificial brain.

The sequence of signals that passes along the centre of the neurons are a genetic algorithm. This is a computer program that evolves to produce the best method of doing something, such as working out the best place to have your fleet of aircraft to ensure minimum delays for your passengers. The algorithm evolves by mutating the sequence of instructions within it.

Once the brain has been wired up the system is set in motion and signals are allowed to flow along the neurons. This sequence can evolve so the brain gets better at what it is being taught to do. The outputs from the brain will be used to control the limbs and actions of the robot kitten.

The big problem is the time it takes to update the sequence of instructions flowing through the network of neurons. Initially de Garis was using Toffoli's CAM machine to crank through these but even this was proving too slow. Now de Garis is going to use specially designed hardware that will be able to update the million neurons in the kitten brain around 2000 times per second. De Garis was hoping to show off the first results by the end of 1998.

Sweet Dreams

While John Conway's Game of Life is a model of the universe in miniature not all CAs are being put to such grand uses. Other researchers are using them to model other dynamic systems such as populations. They hope that what they are discovering will drive the development of societies, helping to avoid conflict and allowing policies to be tried out in an artificial world before they are inflicted on real people.

This chapter opened with a model that has helped resurrect the Anasazi Indians. Behind this re-creation of an ancient culture is a model world called Sugarscape, developed by social scientist Joshua Epstein and computer modeller

Robert Axtell.[29] Both work at the Brookings Institution, a privately funded research organization in Washington, DC.

Sugarscape is not a pure CA. Although the action takes place on a grid of cells, it is not an unbounded space. Their world is small at only 50 cells per side, but this space has not constrained their ambitions and on it they are attempting to model nothing less than the whole of human society.

The landscape grid of Sugarscape is not blank at it is on a typical CA, it is seeded with the little world's only natural resource: sugar. The entities that inhabit this world must find and eat sugar to survive.

These entities are not just cells that are either alive or dead. Each one is an agent, a self-contained program fitted with a variety of abilities and aptitudes. Some characteristics, such as the amount of sugar it owns, health, marital status and cultural identity, can change as the agent moves, migrates and joins with other groups. Other qualities of the agent, for instance eyesight, sex and metabolic rate, are hard-wired and will not change through its lifetime. Not all the entities that fight and die in Sugarscape are created equal. Some are far sighted and can spot sugar from a long way away, others burn the sugar they eat very slowly and can make a small meal last a long time. Other beings are not so well equipped to deal with the rigours of this artificial world, they could be myopic and unable to spot free sugar that is far away, others may have rapacious appetites and swiftly use the sugar they consume to keep them living. If an agent cannot find enough sugar to fuel its search, it dies.

The similarity with other classical CAs that Sugarscape does retain is in its rules. These define the way that agents interact with each other and the environment. There are rules that govern how agents mate, forage and trade as well as rules that govern what happens when they come across some food. The interplay of these rules produces startling results.

The artificial world is set in motion by unleashing hundreds of differently equipped entities on to the grid. Initially organisms head straight for the sugar. In some runs the sugar is piled into two huge heaps, in others it is scattered around the landscape. Those with keen eyesight and a low metabolic rate prosper at the expense of the short-sighted, high metabolic rate organisms. Still other entities that have short sight but low metabolic rate never find the sugar mountains and subsist on the margins eating sugar as it slowly grows back. Axtell and Epstein are happy to say that Sugarscape is a model world because it shares some of the properties of real ecological systems. In the real world habitats possess a carrying capacity, an upper limit on the numbers of organisms they can support. So does Sugarscape. Just as in the real world there are only so many resources to go round and life becomes a struggle to get enough to survive.

Equally important is the fact that the population of agents is never static. It is always fluctuating in response to the amount of available resources and the actions of other agents. A prolonged conflict will deplete numbers, but it also means that there is more sugar for the survivors, who will then prosper. Such boom and bust scenarios are always seen among living populations. That they also appear in Sugarscape lends weight to the claim that it captures some of the essential properties of living systems.

When seasons are introduced the organisms migrate to the most favourable part of the world, marching across the grid like so many miniature nomads following the food.

These simulations have also seen the formation of tribes of agents that share similar characteristics. The changing loyalties of the groups can be studied during times of famine and feast as they travel around the terrain. Axtell and Epstein claim that the behaviour exhibited by the organisms in this simplest iteration of their world closely mimics that found in many hunter-gatherer societies.

All Things Nice

Axtell and Epstein have extended their model to gain an insight into more modern societies. In later models the organisms no longer live in isolation, they breed and are allowed to pass on their stocks of sugar to their offspring.

Introducing this mechanism means that some entities that would die off because of short sight or high metabolism now survive. The downside of this is that some abilities never reach their full potential. The average distance that agents can see across the Sugarscape plain never reaches that found in the harsher world where survival is determined solely by ability. The agents stay a bit myopic. Epstein says that great wealth can easily offset any genetic disadvantages and reduce selection pressure.

In yet another iteration trade is introduced. This is done by giving the Sugarscape world another natural resource: spice. The agents are programmed to need both – if an entity runs out of sugar or spice, it dies – but their metabolic rates are set so they burn them at different rates. Trade is introduced so the artificial organisms do not have to spend all their time hunting wild sugar and spice. When the organisms are made more lifelike by giving them finite lives and evolving preferences for either resource, prices never stabilize and the market swings between highs and lows. Epstein says this shows that traditional economic theory may not reflect the real world. Economics predicts that in a situation where groups have different demands for different resources the market for these goods should gradually move to equilibrium so the benefit for everyone is maximized. Yet Axtell and Epstein have shown that this is not what happens. Axtell says that the assumption that a market economy is inherently efficient may well be false and that perhaps we need a better way of sharing around the resources in our world.

Now the researchers want to move on to situations that map directly on to real world concerns. One idea is to mount a study of international relations by setting up a series of small nations and letting them interact. The two colleagues are interested to see if they can re-create the Cold War, the formation of NATO or even a world war.

Axtell, Epstein and others have not ignored the possibility that a study of Sugarscape might reveal the fundamental flaws in the way Western society is organized. Sugarscape is part of a larger endeavour that is known as Project 2050. This is an ambitious research program that is attempting to identify the conditions for sustainable development on a global scale. The policy-makers and researchers involved in the project believe that modelling artificial societies can, and should, shape what we do in the real world.

Axtell and Epstein are now working to extend Sugarscape so that it comes to capture the way of life of people in the late twentieth century. So far the agents in Sugarscape have sex but there are no families, no cities, no companies and no government. Over the next two to three years they hope to be able to produce a model in which all of these things emerge spontaneously. There is also the disquieting possibility that social structures such as cities and governments have not emerged because they are not needed in a world as simple as Sugarscape.

While the work they have done so far is suggestive, Axtell is wary of using Sugarscape to form policy or run local governments. It is not quite sophisticated enough for that yet. 'We have been quick to point out that this approach is in its infancy,' says Axtell, but he says that one day it should be able to help and inform policy decisions. At the very least it will give administrators the chance to test their policies before they are implemented to see who wins, who loses and whether the plan achieves the aim the policy-makers want it to. They can run as many experiments as they want in Sugarscape without hurting anyone, a choice that they have not had before.

Sugarscape is an important model for several reasons. Firstly, although the social sciences study society they do so in isolation. Economists, geographers and sociologists rarely pool the knowledge they have. The organization of departments in universities and colleges reflects this divide. Sugarscape brings everything together. All the insights are thrown into the world and tested at once. With Sugarscape social scientists get a chance to study how their fields interact and how each one contributes.

Secondly, Sugarscape is dynamic, unlike the social sciences, which tend to deal with single instances rather than the flow of behaviours that emerge in a society. It may not be easier to hit a moving target but at least Sugarscape lets social scientists follow its tracks.

Thirdly, Sugarscape preserves the differences of culture

and aptitude found in humans. Usually these are smoothed away or ignored in sociological analysis.

Finally, Sugarscape gives social scientists the freedom that colleagues in so-called hard sciences[30] have had for a long time, the ability to repeat experiments and test hypotheses. For the first time social scientists have a test bed for their theories. They can prime Sugarscape with the problem they want to study and watch it run and see what matters and work out just what caused a particular social phenomenon to emerge or disappear.

Honest Indians

Although Axtell is wary of using Sugarscape to form and test public policy it has already been used to study simpler societies, in particular the Anasazi Indians. A taster of some of this work opened this chapter. George Gumerman says they picked the Anasazi because they are such a mystery and because there is a lot of information about their culture through most of its existence, largely because when they abandoned their villages they left a lot of artefacts behind. They didn't try and model the whole Anasazi tribe, instead they picked a small but significant settlement centred on Long House Valley in the north-eastern corner of Arizona.

The decision about which site to study was aided by the historical data available. To make the model realistic they needed to be able to reproduce conditions as closely as possible, especially because they suspected that the reason the Anasazi upped and left was because of a change in environmental conditions. Jeff Dean has spent the past thirty years collecting an enormous amount of data about climatic conditions in Long House Valley. He has compiled rainfall records going back thousands of years by studying tree rings. These are good guides to ancient weather trends simply because in wet years tree rings are thicker. Dean has also collected detailed information about the way soil and vegetation have changed over the centuries.

All this detailed information was used to make the landscape on which the artificial Anasazi live. Also on this landscape are the sites of other Anasazi settlements and the fields where they raised crops. The Indians are modelled down to the level of individual households. The households are even matrilineal like the Anasazi. They farm their land just like the real Indians. The figures Gumerman and his colleagues have on yields per acre have helped refine this part of the model. Says Gumerman: 'This is not a simulation, it is a reconstruction.' To all intents and purposes the Anasazi are living again.

As the description of the computer-generated Anasazi at the beginning of the chapter makes plain, Gumerman and his colleagues have not got it right yet. In their computerized valley the Anasazi do not migrate away like the original Indians did, but the researchers are tantalizingly close.

Over twenty-five years ago Gumerman predicted that a change in environmental conditions led the Anasazi to abandon some farmland and start thinking about leaving their settlements. Now they know that the truth is more complicated than that. They now know that environmental factors did play a large part in driving down crop yields and driving away the Anasazi. 'We know [the exodus] was triggered by environmental change,' says Gumerman. 'There was a slight drop in the water table, erosion set in and precipitation became more spotty.' But, he adds, this was not the whole story.

When these environmental conditions were re-run in the Sugarscape model the numbers of Indians dwindled but not to the numbers that were thought to exist at the time. Also far fewer Indians abandoned the area altogether.

Gumerman thinks that social factors contributed to the Anasazi's decision to leave Long House Valley and head south. The latest work by the team is changing the social structure to add clans – early runs have shown that this drives down numbers further – but the Indians are still surviving in large numbers. A piece of the puzzle is still missing.

Sugarscape has its critics and many are not convinced it offers anything useful. 'The modellers are moaning that we are making it too complicated and the archaeologists are moaning that it is not complicated enough,' says Gumerman ruefully. They must be doing something right to annoy both groups. Gumerman is excited by the possibilities that computer models like Sugarscape offer social scientists. In the past it was difficult to decide which of any competing theory was closest to the truth. Sugarscape gives social scientists a chance to run different explanations through the model and see how they work out. The skill is in identifying the most important influences. 'Stephen Jay Gould says that we cannot re-run the tape,' Gumerman says, 'but that is exactly what we are doing.'

Join the Dots

CAs may also be the ideal method of making computers smaller and even more powerful than they are today. While computer companies are managing to cram ever more transistors into microprocessors they are approaching several fundamental limits that will be difficult, if not impossible, to overcome.

Currently the smallest components on a silicon chip are around 0.35 microns across. While chip makers confidently expect to be able to go beyond this their confident predictions run out below the 0.1 micron level. The reason for this is that at this scale the components that make up a chip will be so small that quantum effects will start to make themselves felt. Transistors that were reliable when 0.2 microns across will become unpredictable at 0.1 microns. Building a microprocessor out of components that do not act like you expect them to will scupper any attempts to work on such small scales.

Then there is the problem of how to make the components in the first place. At the moment the successive layers of a silicon chip are etched out of a substrate by shining light through a stencil to create a pattern. This is then etched away using a variety of corrosive chemicals, leaving the desired design on the chip. But to etch components at sizes below 0.1 microns means using ultraviolet light or X-rays. Visible light is too blunt a tool to use to cut components on the sub 0.1 micron scale. Changing to ultraviolet or X-rays is no trivial task when fabrication plants cost many billions of dollars to set up and nearly as much to re-tool. While research into using X-rays and ultraviolet to etch components is proceeding apace, commercial systems are unlikely to be ready by the time computer chip makers need them.

Finally there are the problems associated with working with such small components. The more transistors there are crammed on to a square of silicon the more heat they produce and the harder it is to keep them within their working temperature. Some of Intel's Pentium chips have to have a cooling fan mounted directly over them to prevent them overheating and stopping working.

At sub 0.1 micron scales the distance between components on opposite sides of the chip becomes significant and it gets harder and harder to synchronize what every bit of the chip is doing. Now some scientists are suggesting a radical departure from the existing method of making chips. What is attractive about this approach is that as the individual elements of the processor shrink, performance improves. These scientists want to use devices known as quantum cellular automata (QCA) to process information.[31]

Existing chips predominantly use transistors as switches to charge and discharge capacitors to specific voltages that represent the underlying logic. This is how they store or process data. But in a QCA logical states are encoded by the positions of individual electrons. In this respect QCAs are very similar to ordinary CAs in which it is the arrangement of 'live' cells that determines what happens next. QCA researcher Alexi Orlov of Notre Dame University in Indiana says it would be difficult but not impossible to use QCAs to explore issues of artificial intelligence and artificial life. For himself, though, he is more interested in the benefits to processing that a QCA can bring.

The basic element of the QCA is an arrangement of four tiny aluminium islands, or dots, interconnected by what are known as tunnel junctions. Two of the dots act as inputs, the other two as outputs. By manipulating the polarization of the dots, electrons can be moved through the junctions and forced to exit via either output dot or stay put. In this way it is possible to mimic all the operations of Boolean logic used in existing computers.

The advantages of the approach are that the tiny dots are made the same way that existing chips are so they should be easy to adopt as minimal re-tooling of a fabrication plant will be needed. The main disadvantage at the moment is that the device only works at very low temperatures, too cold for commercial use. But now the researchers are working on shrinking the whole device to make each dot out of individual molecules. When the dots are this small their relative energy will be much higher so the device should be much more stable and work at room temperature.

Chapter Three

Rise of the Robots

The mass of humans are born slave-drivers. Just look at the Asimov priorities: Protect Humans. Obey Humans. Protect Yourself.

Humans first and robots last? Forget it! No way!

Rudy Rucker, *Software*[1]

Espionage and Experiments

Science and secrets are intimately bound up with the origins of artificial life. Some of the twentieth century's greatest scientists, and a few of its most successful spies, were educated at Cambridge between the two world wars. In 1931 two men who would go on to complete work without which artificial intelligence and ALife would never have got started began their studies at the university. Without this solid scientific background many of the robots ALife experimentalists have subsequently made and the software worlds they have created that today are re-writing the rules would be laughed out of the laboratory.

The two men who were so influential are Alan Mathison Turing and William Grey Walter. Today Turing is as well known for the breakthroughs he made in mathematics and the work he did on cracking German codes during the Second World War as he is for his contribution to AI. Walter, on the other hand, is known to few outside ALife. His work is often referred to in passing but rarely fully explored. Even within neurophysiology, a field in which he made equally fundamental discoveries, his name does not engender the respect it perhaps should.

When they began studying no one could have guessed what Turing and Walter would go on to accomplish. Of the two, Turing was the one who would shine first, completing his most lasting contribution while still an undergraduate, though it would take years for its full significance to become obvious. Walter made his mark after he graduated and initially gained more notoriety for it. Shortly before a tragic accident cut short his academic career, Walter was a TV pundit and a novelist and his early experiments with artificial lifeforms were written up in the national press. The effect both of them have had forty and fifty years later is profound – few ALife papers are written that do not owe a debt of some sort, acknowledged or not, to Turing and Walter.

Although today the work that these two men completed is intimately associated with ALife and addresses the same problems, it is unlikely that they met during their student days. After they graduated there is plenty of evidence to put them in the same place at the same time. Their interests in later life began to coincide and by the 1950s they probably knew each other socially. As students, however, they moved in very different social circles that rarely, if ever, mixed. Walter was friend and confidante to the privileged elite of the university who forty years later would be exposed as spies and traitors. He never turned his back on these men or the beliefs they all shared. Turing attended lectures with some of the same men that Walter knew but had little or nothing to do with them outside the classroom. He did not share their faith and thought them frivolous. In turn they thought him provincial and dull. Like them Turing had secrets to hide, but his shyness prevented him from seeking out those with similar burdens. In the end, when his secrets were revealed Turing killed himself.

Good Science

Turing and Walter could not have chosen a better place or time to start their scientific careers than Cambridge in the 1930s. At the time it and the University of Gottingen in Germany were the two scientific centres of the world.[2] Cambridge owed its reputation to a series of breakthroughs in physics made at its Cavendish Laboratory during the late nineteenth and early twentieth centuries. Much of the early work on quantum physics had been done at Gottingen and it was here that renowned mathematicians such as David Hilbert worked. In the 1930s Cambridge was still in the ascendant while the ability of Gottingen to maintain its reputation was coming under threat. The turmoil Hitler created once he had seized power forced many of its best scientists to flee to the USA or Britain to continue their work.

The changes that Hitler's Germany would bring lay in the future when Turing and Walter went up to Cambridge. At that time the German university could still boast a formidable line-up of researchers. Still working there were men such as Max Born, James Franck, Werner Heisenberg and Max von Laue, names synonymous with the new field of quantum physics.

But the Cavendish also had its stars. At the time that Turing and Walter went up to Cambridge Ernest Rutherford had been running the Cavendish for ten years. He took over from Joseph John Thomson, who discovered the electron, in 1897 – two years after Rutherford started working at the lab. An assessment of the effect of Thomson's reign concluded that he 'more than any other man was responsible for the fundamental change in outlook which distinguishes the physics of this century from that of the last'.[3] Thomson was instrumental in showing that Lord Kelvin was very wrong to declare at the turn of the nineteenth century that physics was nearly completed and all that remained to sort out were the details.

Even before he took over in 1919 Rutherford had established a reputation both as an innovative experimenter and scientist. During this twenty-four-year period, nearly all of which were spent away from Cambridge, he carried out pioneering work on nuclear energy, established that the atom was not a solid particle, followed this up with the discovery of the proton and with Hans Geiger invented the radiation counter that bears the German's name.

In 1917 he completed the experiment that most people know him for. In the words of the world's popular press he 'split the atom'. Rutherford bombarded nitrogen atoms with alpha particles to produce an artificial nuclear reaction. Four years after this Rutherford delivered a speech at the annual meeting of the British Association for the Advancement of Science which in that year was held in Liverpool. During his oration he said: 'We are living in the heroic age of physics.'[4]

What Rutherford neglected to mention was that he was one of the heroes.

During his time as head of the Cavendish Rutherford recruited scientists who would maintain the reputation of the lab in the face of growing competition from the US. Other historians have noted: '. . . the team that Rutherford built around him in the twenties . . . contained the largest number of gifted experimental physicists grouped together which had ever been known.'[5]

In mathematics too the university had an enviable reputation. Although the early work on quantum physics had been done in Germany, much of the mathematics needed to make it work had been done in Cambridge, and the lecturers who schooled Turing were world authorities in their subjects. Prior to going to Cambridge Turing had been reading popular treatments of breakthroughs in the sciences written by the very men he was now being taught by.

The scientific world in the opening years of this century was a much smaller one than it is today, and one that Cambridge managed to dominate. All it used to take to establish a world-beating reputation was a few gifted, busy men. In his panoramic history of the twentieth century *The Age of Extremes*, Eric Hobsbawm estimates that in 1910 there were in total only 8000 German and British physicists – nearly enough to fill a large village.[6] In such a small crowd it was easy to know the big names – many of them would probably be your tutors – and be on nodding terms with many more. The scientific community was exactly that. By the late 1980s Hobsbawm writes that the number of practising scientists had grown to over five million – as many people as live in inner London.[7] In a crowd this size it is much harder to be noticed.

In the 1920s although the number of those engaged in science was growing, the depredations of the First World War meant the rate of growth was slow. The war killed off many good scientists and more who promised great things. Because of this the number of people engaged in science stayed largely static. The community was still small and easy to know one's way around. When Turing began studying mathematics he was one of only eighty-six scholars to take up the subject that year.[8] The college he joined, King's, had only sixty students in each year.[9] Kings and Trinity were the two colleges that the sciences drew most of their students from. Walter was also a King's man and Rutherford was from Trinity.

Proof Positive

It did not take Turing long to show that he was capable of living up to the reputation of King's and Cambridge. Turing graduated in 1934 with a distinction – one of only eight to do so; with this came an award of £200 per annum for a research studentship. This was enough money to live on while he tried for a fellowship at King's.[10] He finished the dissertation quickly. It was a rediscovery of a theorem first put forward by mathematician J.W. Lindeberg in 1922 that Turing had proved independently. He handed in the paper a month early. It was enough. In the spring of 1935 he became the first King's fellow of the year. As a fellow of the college he got more money – £300 per annum – and had three years with no teaching obligations in which to research anything that interested him.

Turing picked a subject that not only interested him but

also was on the minds of nearly every mathematician on the planet at the time – that of putting their subject on a firm foundation.

In 1928, at an international mathematical congress, David Hilbert had challenged mathematicians to prove that their subject was more than just an artificial creation. The task he gave them was to prove that, although mathematics dealt in abstractions like numbers and could give rise to some paradoxes, it had at bottom a solid basis. He challenged them formally to prove that mathematics was more than thoughts, to investigate its underpinnings and prove that it all really did make sense. Formal in this sense means establishing the truth of a proposition using the rules of logic. He was asking them to establish once and for all that there were no situations in mathematics that would give rise to nonsensical answers. Essentially he wanted to be sure that 2+2 could never equal 5.

Hilbert was the foremost mathematical thinker of his day and in setting this challenge he was responding to the sense of crisis felt by many mathematicians at the time. This feeling had been brought about as maths had become more abstract and in doing so had thrown up paradoxes that left many feeling uneasy about the subject.

To remedy this problem Hilbert set very specific goals. He challenged mathematicians to prove their subject fulfilled three separate conditions. That it was complete, consistent and decidable. Logic assigns very precise qualities to each of these terms. In a complete system of mathematics (or logic) there are no unprovable statements, for every assertion such as 'every integer is the sum of four squares' there should be a way of establishing that it is true or false. The method of establishing the truth or falsehood does not have to be the same in every case just as long as there is a way to do this. A consistent mathematical system is one that cannot produce false results using valid rules. So it should be impossible to reach conclusions like 2+2=5 as long as valid rules are followed. Finally he asked them to establish the decidability of mathematics. This is to do with finding a method that could be applied to any and every assertion or statement that would show whether the assertion was true or false.

Unfortunately at the same meeting an Austrian mathematician called Kurt Gödel dealt this project a mortal blow. To do this Godel used the system of logic espoused in the 1910 three-volume opus *Principia Mathematica* written by Bertrand Russell and Alfred North Whitehead. This had tried to show that mathematics did indeed have consistent foundations but in a roundabout and, some thought, overly complex way. Using this system of logic Gödel was able to produce a proposition that was unprovable, and in doing so he showed that the Russell–Whitehead system of logic was inconsistent and incomplete and therefore that it was a system that could not save mathematics. To do this he showed that in the Russell–Whitehead system it is possible to produce a mathematical proposition that is the equivalent of: 'This statement is unprovable.' Although this statement is true it is impossible to prove it so. If it is unprovable, it is true, but true statements can be proved and this one, in its own words, cannot. Although you and I can appreciate that the statement in question is true, or at least bat it back and forth in our minds until we are convinced or bored, there is no axiomatic way to establish this truth.

This was the exact opposite of what Hilbert was looking for and seemed to spell the end of his grand scheme. However, there were other logical schemes in existence,

Hungarian wunderkind John von Neumann had invented one. Could this, or a different one, do the job better than the Russell–Whitehead scheme?

Unfortunately Gödel demonstrated the shortcomings of the Russell–Whitehead system in a way that applied to any and all logical systems. He established, proved, that in any formal system it is always going to be possible to produce statements that are unprovable by that system. Establishing the truth or falsehood of these statements required one to mentally step outside that system and consider them afresh. This therefore meant that it is impossible to come up with a system that is consistent and complete, as a consequence trying to establish the foundations of mathematics on a logical basis is doomed to failure. Hilbert's dream lay in ruins, there was no way to show that mathematics was more than just a convenient shorthand for dealing with some of the trickier aspects of the world.

Thankfully this was not the end of it, there was still hope. Gödel had only shown that maths cannot ever be consistent and complete. This left the final challenge. Mathematics may still be decidable. There may exist a finite number of steps that could be applied to mathematical statements to show that individually they met the standards Hilbert hoped they did. Working through them using this method might be a long-winded way of giving Hilbert what he wanted.

It was this third challenge that Turing decided to take up. Prompted by a phrase of one of his lecturers, M.H.A. Newman, Turing set out to find out if there was a mechanical method of proving mathematics decidable.[11]

Man and Machine

Intellectually Turing was of a very practical frame of mind; as a teenager he had designed and attempted to build his own typewriter,[12] so it is no surprise that he came up with a machine to try and do what Hilbert wanted.

In the early summer of 1935 while resting in a meadow near Grantchester after completing one of the afternoon runs that he had started to take, Turing was struck by a flash of inspiration – a way of establishing the decidability of mathematics.

Turing conceived of a very special type of machine that could manipulate the symbols presented to it and perform the necessary operations on them to test them for their truth. This lean, stripped-down minimum machine would automatically go through those mathematical statements, interpret the symbols they were written in, establish their truth or falsehood and move on.

The Turing machine is entirely abstract and was never meant to be built, but he imagined it resembling an old-fashioned stamping mill. Instead of a mould that would be pressed into a sheet of metal passing beneath it, the Turing machine had a scanner or reading head that was capable of reading a tape that passed beneath it. This tape was divided into squares and each square was either blank or contained a symbol. The symbols could be numbers or letters but they are most often conceived as either zeroes or ones – binary digits. The tape is fed beneath the reading head and once it has read the symbol it sees there the machine takes action. The head can be in any one of a number of configurations or states. Depending on what state it was in and the symbol it was contemplating it would perform one of a finite number of actions. It might, for example, erase the symbol and leave the previously occupied square blank, replace the old symbol with a new one, leave the square untouched and move on, stay in the same configuration or change to a new state, move on to the next space to the left or right or stay in the same position.

The machine knew what to do in every state – this information was written up in a table that the machine worked to. The table could represent any set of rules that you wanted it to, including those of formal logic. Using these rules and these few simple actions the machine would be able to test mathematical propositions. Turing realized that because it is possible to produce unprovable statements using the same rules there can be no method of sorting the good from the bad. The rules cannot be bent but they can produce statements that cannot be proved.

The title of the paper, published in 1937, that contained the details of the Turing machine was 'On Computable Numbers, with an application to the *Entscheidungsproblem*'. The *Entscheidungsproblem* was the third of the challenges that Hilbert posed. The reason that mention of it is relegated to the end of the paper's title is because by the time that Turing finished the paper he saw that the machine he conceived of had a much wider relevance.

Turing called his machine the universal machine because it was capable of emulating the workings of any other machine, although it would undoubtedly take much longer to do the same job. In fact all other Turing machines were really more sophisticated universal machines that just did the job faster.

Turing considered the human mind to be a machine,

albeit a fantastically complex one, and as such ripe for emulation by a universal machine. All that was needed was the table that told the machine what to do in all circumstances.

In 'On Computable Numbers' he 'gave some pages of discussion that were the most unusual ever offered in a mathematical paper'.[13] These dealt with the parallels between human computation and that of a machine. In essence what a Turing machine was doing when it was performing its scanning, reading and writing was manipulating symbols. At the time this was thought to be central to human thought and consciousness. As a hard-headed materialist Turing reasoned that there is nothing within the brain that is not the result of interactions of smaller elements. In a sense everything is determined and because there are finite numbers of chemicals crashing about there are only so many 'states of mind' that a brain can be in. There may be a lot of them, a practical infinity perhaps, but a universal Turing machine should be able to emulate the workings of the human mind albeit on a crude and slow scale. It might be possible, therefore, to capture the essence of humanity in a machine. A less specialized machine than Turing's universal machine may take less time to grind through all the states of a mind and may prove to be a better representation of conscious thought.

It is in this latter sense that Turing's work is most important. At the time the paper was finished its impact was diluted because just before it was published Turing discovered that another mathematician, Alonzo Church, had come up with a similar way of showing that there was no method of deciding whether a mathematical statement was true or false. Turing's lasting contribution is the machine that bears his name and the bridge it erects between the worlds of mind and machine.

The only flaw with Turing's insight, that it might be possible to make a conscious computer, is that it too may be subject to the shortcomings Gödel uncovered. It may be that there are states of the mind that are uncomputable by that system. Roger Penrose in his books *The Emperor's New Mind* and *Shadows of the Mind* argues the case for some aspects of consciousness being uncomputable. Certainly it seems to take something that computers do not possess and perhaps minds do to recognize that 'This statement is unprovable' is what it says it is. Penrose claims that what is needed to make sense of mind and consciousness and by-the-by unify gravity with quantum theory, is nothing less than a rewriting of physics. Even if this supposition is true, and there are many convinced that it is not,[14] then ALife may show a way out of it, for ALife is as much about breeding as it is about building.

Computer Creator

The significance of 'On Computable Numbers' did not become immediately obvious, but the novelty of its 'shockingly industrial'[15] solution did establish Turing as an innovative, original thinker. It also helped him win a scholarship to Princeton, where he met and worked for a while with John von Neumann, bringing the ideas in 'On Computable Numbers' to the attention of the great man. In fact von Neumann was so impressed with Turing that he wrote a letter to Cambridge supporting Turing's nomination for a scholarship that would enable him to stay at Princeton for another year.[16]

After two years at Princeton Turing had had enough of America and when his fellowship of King's was renewed in March 1938 he returned to Cambridge, this despite being offered a job at the Institute of Advanced Study in Princeton. At the time the IAS had a very short faculty list, but then there were few people that could compete with men like Einstein and von Neumann. Shortly after Turing turned down the offer of a job, the IAS was joined by one of many scientists fleeing Nazi Germany. He stayed thirty years until mental instability forced him into a sanatorium. The man was Kurt Gödel.

As the world prepared for war Turing too became embroiled in the conflict. For many years he had been interested in codes and ciphers. As a schoolboy he had infected a friend with his interest in secret ways of communicating and regularly sent the friend messages in the form of strips of paper with holes punched in them. His friend would then have the onerous task of plodding through the book supplied with the paper and finding on which page the holes revealed a message.[17] At Princeton he had designed and built a working electronic multiplier that could be used to produce uncrackable codes. This travelled with him on the boat back to Britain when he returned in 1938.

Soon after he returned he was recruited to help the Government Code and Cypher School at Bletchley Park, probably on the recommendation of one of his tutors who had worked at the same place when it was established during the First World War. In 1938 the school was gearing up for war and had as its task the cracking of German codes, specifically working out how the German Enigma machine scrambled information. Variants of the Enigma were used by nearly every branch of the German military and working out how it operated would give the Allies a significant advantage. The school had limited success unravelling the secrets of the Enigma machine until 1939. In July of that year Poland made Britain a gift of information about the Enigma machine that Polish mathematicians had gleaned from years of working on the problem.

When war broke out in September, Turing started working full time at Bletchley and began looking for ways to make better use of the Polish intelligence. The Poles had already designed a machine called the Bombe to take the drudgery out of checking through the hundreds of possible permutations of a message. Working with Gordon Welchman, Turing redesigned the machine to make it a much more powerful device, then working alone he produced a more powerful version that worked twenty-six times faster. Using this refined Bombe the Allies were able to decipher Luftwaffe signals as they came in.[18]

Turing then turned his attention to the system used for German naval communications. Although this service also employed the Enigma machine, it did so in a much more complex fashion. Turing managed to work out how to break this code by 1939, but it would take another two years, the capture of vital information from German U-boats and the development of new statistical methods to ensure messages could be read regularly. The advantage was held only briefly because in February 1942 the U-boats changed the way they encoded messages, negating all the progress that Turing had made.

In a bid to make the decoding Bombe machines work

even faster Turing began working with telephone engineers on ways to adapt electronics to the task of creating the Colossus machine that would crack the system used to encode Hitler's strategic communications. His work at Bletchley was interrupted briefly in late 1942 when he was sent to the US to exchange information about the problems of deciphering the messages U-boats were using. While working at Bell he met information theory pioneer Claude Shannon for the first time and found he was interested in the same problems that Turing was. Prior to his return to Britain in March 1943 the U-boat problem was solved by the discovery of logical weaknesses in the Enigma encryption scheme. For the rest of the war the Allies could once again find out what the U-boat commanders were being ordered to do.

Back at Bletchley Turing oversaw the construction of the Colossus machines that were turned on just before D-Day. As well as being a breakthrough in cryptographic methods Colossus was the first large-scale digital computer. It was the forerunner of all modern computers. At the same time Turing was toying with the idea of building a working universal machine of the type he had imagined in 'On Computable Numbers' in 1936. Only now, fifty years after the end of the Second World War, is it emerging how far ahead of its time the work at Bletchley had gone. At the peak of activity there were over 8000 people working on decoding German signals. Churchill considered the Colossus machines to be so sensitive that at the end of the war they were taken away and the blueprints burnt. Tony Sale, who worked at Bletchley during the war, has spent the last few years trying to rebuild some of these machines. He was stymied until 1996, when the Government finally declassified the files relating to Colossus. Even then GCHQ – the modern equivalent of Bletchley – would only let him build one of the earlier models of Colossus that had no memory.[19]

One result of Turing's involvement with the computers at Bletchley was that he was asked to help academics at Manchester work on their design for just such a device. In 1948 a small prototype computer known as the 'Baby' became the first working modern computer. It was the model for a computer called the Manchester Mark I, which was modern in the sense that it stored both its program and the data that the program would use in the same memory. This made the computer much more flexible; turning it to a new task meant simply loading a new program and data. Previously every computer had to be physically rewired every time its users wanted it to attempt a novel task. Although von Neumann, Eckert and Mauchly had come up with the same idea in the design for the EDVAC it would be years before that machine was completed. The Manchester academics beat them to it.

Throughout his life Alan Turing was searching for a way to penetrate the mysteries of the mind and find a way to reproduce what it was that was going on inside people's heads. Not only did he single-handedly establish a way of using machines to do this in 'On Computable Numbers', but he was also intimately associated with the creation of the first computers that could make the building of the machine postulated in that paper a reality. More powerful computers would be used by the founders of artificial intelligence research to establish their field. Unfortunately the power these computers seemed to put into the hands of these men led them to make rash claims for their chosen subject and promise far more than they could deliver.

The Seduction of the Innocent

It was easy to see how those who were used to working everything out with paper and pencil and the occasional mechanical calculator could be seduced by digital computers. The first computer built at Manchester had a memory of only 2048 digits, but its successors rapidly improved on this. In 1951 Ferranti sold the first commercial computers based on the Manchester design and with a memory counted in tens of thousands of digits.

Soon smaller, more powerful versions were being built using transistors and sold by the English Electric Company. As one AI historian puts it: 'Early progress in using computers for arithmetic were truly breathtaking. In a few years, technology went from cranky mechanical calculators to machines that could perform thousands of operations per second. It was thus not unreasonable to expect similar progress in using computers as manipulators of symbols to imitate human reasoning.'[20]

Turing's universal machine captured the essential belief of anyone working on AI in its early days. They believed, as fervently as Turing did, that every facet of the mind, everything from learning to vision to composing music, can be dissected and described in such detail that a machine can be made to simulate the process. To these people, and many that have followed them into AI, the mind has no direct access to the world, instead it uses an internal representation constructed using symbols. If the mind uses symbols to think with, then any other device, say a computer, that does the same thing should display the same properties as a mind.

Certainly the pioneers of AI, men such as John McCarthy,

Marvin Minsky, Herbert Simon and Allen Newell, were seduced by the ever-increasing power of computers.

In 1956 McCarthy and Minksy organized a conference in Dartmouth, Massachusetts, that attempted to bring together all those working in the new field of AI. This conference was also the first place that the phrase 'artificial intelligence' was uttered, apparently by McCarthy, and it is where the community first got to know each other. Over the next twenty years all the major developments in AI would be completed by the people who attended this conference, or their students.

Nearly every person attending this two-month conference was working with computers and looking for ways to use them to mimic the mind. But only two attendees, Newell and Simon, had a working program to show off. The program was called Logic Theorist and as its name implied it was an automatic method of solving some of the propositional calculus theorems of Russell and Whitehead's *Principia Mathematica* – a nod to the problems Turing was trying to tackle with his universal machine. Logic Theorist eventually managed to prove 38 of the first 52 theorems in Chapter Two of the Russell and Whitehead work. One of the proofs it came up with was even more elegant than that derived by the two authors.

Newell and Simon persisted with this approach for the next twelve years. One of the most successful programs they devised was known as the General Problem Solver (GPS). It was modelled on other work that mimicked the way that humans go about solving problems. GPS eventually learned to solve a variety of puzzles, broke secret codes and performed some symbolic integration.

Others followed up this continuing success. Herbert

Gelertner at IBM created a geometry theorem prover in the late 1950s. It was significant because it worked on an internal representation of a geometrical figure fed into the computer using punched cards. Others at the same company wrote programs to play draughts (checkers) and chess. These latter two programs were reported up in the national press. Work on them stopped shortly after IBM chairman Thomas Watson was asked in a shareholders meeting why the company was engaging in such frivolous research.

AI received a boost in 1962 when MIT received a multi-million dollar grant from the Advanced Research Projects Agency (ARPA). The money was to be used to 'research new ways in which computers can help people in their creative works, ranging from research to education and management'. Using this money Minsky, who by now was working at MIT, set up the Machine Aided Cognition program.

The hubris of the researchers working on this project became obvious with the problem they chose to tackle first: vision. In 1966 Minsky and his colleagues thought they could figure out how vision worked in a year and gave the job to graduate student Gerald Sussman to work on. Today few people would claim to know how it is that we see and interpret the world around us.

When Sussman failed to crack vision over the summer of 1966, Minsky and others at MIT realized that the problem might be more difficult than they first supposed. To simplify the problem they turned away from the real world and created an artificial environment in which to test their ideas. The Micro Blocks World that resulted was a featureless box populated with a few simple geometric shapes.

Most of the work on the Blocks World was carried out

by Seymour Papert – the son of an itinerant South African entomologist. His work showed that in simplified worlds it was possible to make a computer interpret what it saw, manipulate the blocks and even answer questions about them.

Another MIT worker, David Waltz, was working on the Blocks World and he built a coupled robot and computer system that could recognize what it saw, say an arch made up of two uprights and a lintel, and build a copy. It knew how to recognize near misses too.

The success of this led to the work being copied at Stanford, where John McCarthy was now running the AI department. The Stanford team built a robot called Shakey that existed in a life-sized Blocks World. This was made up of seven rooms connected by eight doors. In some rooms were square boxes that Shakey was supposed to push around and stack into shapes. Instructions were given to Shakey via a keyboard and then the robot was supposed to navigate its way around the rooms and do what it was told, recognizing what it wanted when it saw it, collecting the boxes it needed and stacking them up. The cash needed to build Shakey came from the military via ARPA, which thought that if the project was successful it would get a mechanical spy it could send behind enemy lines to reconnoitre.

As might be gleaned from its name, Shakey was not a huge success. When it moved around it shook and rattled and it had other failings too. 'The shakiness of the robot was not its major weakness; its inability to recognize whether it had successfully completed a task was. If Shakey failed to make its way through a door before encountering a stack of blocks, it would move about as if it had gone through the door and attempt the block-stacking procedure anyway.'[21]

Two versions of Shakey were built, one in 1969 and one in 1972. But after five years of pumping money into the project ARPA (which had now become DARPA) was looking for more than what it got – a shaking, stupid robot that did not even know where it was. DARPA cancelled the funding for Shakey.

Back at MIT some progress was being made with a program called SHRDLU created by Seymour Papert. This program could be asked questions about the Blocks World and gave answers in conversational English. To get it to do this meant sacrificing some of the formalities of programming languages but the results (for the time) were so spectacular that this seemed a reasonable price to pay.

Unfortunately SHRDLU was the only real success of the MIT Blocks World project. At about this time, the early 1970s, the failures of the early AI projects mounted up. Research grants into machine translation were cut off because results were so poor. Attempts to expand what had been learnt in the Blocks World to the real world failed abysmally. At the end of it all many AI researchers were left with a few stupid robots and no real insights into consciousness apart from the realization that it was much harder to capture than they had previously thought. 'The sad truth, as proponents of the micro worlds realized, was that you cannot define even the most innocuous and specialized aspect of human usage without reference to the whole of human culture. The techniques used in SHRDLU would not work beyond artificially defined toy problems or restricted areas of expertise. Contrary to expectations, the micro worlds approach did not lead to a gradual solution of the general problem of intelligence.'[22]

In fact the way forward had already been found nearly twenty years before mainstream AI researchers realized they were on a hiding to nothing. The researchers had missed it, largely because they were so impressed with computers and thought the problem would be solved by throwing more computer power at it. One man had not missed it, however, and that man was William Grey Walter. The story of the success that his work had brought about begins in Cambridge but ends on Mars.

The Perfect Spy

The privileged elite at Cambridge may not have sought out Alan Turing's company but they extended a much warmer welcome to William Grey Walter. This despite the fact that Turing had a solidly British upbringing and Walter started his life in America.

Walter was born in Kansas City in 1910, the only child of American Margaret Hardy and Englishman Karl Walter. At the time, Karl was city editor of the local newspaper, the *Kansas City Star*. Despite this local success, Karl was not wedded to Kansas or America.

When the First World War broke out he left his wife and child safely behind to go and serve his country. He did not go to the front line but the work he did was no less vital. According to Nicolas Walter, one of Grey Walter's surviving children, Karl spent the war working for British Intelligence.

His job was to try and entice the Americans into the conflict by feeding them information that was likely to sway their judgement.

Whether the work he, and others, carried out was successful or not is hard to say. It is enough to note that America did join the war effort as an 'associated power' in April 1917. Later that same year Karl returned to the US to collect his wife and son and bring them back to Britain.

Once in Britain Walter embarked on an archetypal English education. He attended a Bayswater preparatory school until 1922 and then, when he was 12, he entered Westminster School, a very proper English day school that can boast as old boys Ben Jonson, John Gielgud and Peter Ustinov. While there he spent most of his time studying the Classics (Latin and Greek). He excelled at these subjects and while at Westminster won a distinction in Greek and as a result was recommended for a classical scholarship to Oxford.

This commitment to the classics was not due to a love for the subject. There was another reason for his devotion. At home and at the family's weekend cottage in Surrey Walter had been conducting his own course of study, not more classics but science experiments. This abiding interest in all things scientific led Karl to promise Grey all the scientific apparatus he wanted provided he stuck at the classics.

During his adolescence Walter developed a passionate and private interest in chemistry, microscopy, biology, astronomy and physics. Like many other pioneering robot makers he was an experimenter, gadget maker and tinkerer from an early age, amusing himself by making his own radio as well as other gadgets. This interest in electronics stayed with him all his life. Many of the advances he made in neuroscience were due to his invention of devices to study different sorts of brain waves.

As his interest in science grew Walter's devotion to the classics waned. As soon as he got the chance to drop the subject he took it. He rejected the chance of studying Latin and Greek at Oxford in favour of natural sciences at Cambridge. In 1928 he entered the university as a scholar of King's College. The wisdom of his choice and his devotion to science were revealed three years later when he was awarded a first-class degree in natural sciences.

He stayed on at Cambridge to do postgraduate work and spent four years studying the physiology of nerves and conditioned reflexes. Through this work he met Pavlov, the Russian psychologist who had pioneered work on conditioning the responses of animals, famously making a dog salivate when a bell was rung by conditioning it to expect food at the sound of a bell. This work kindled another life-long interest of Walter's, his fascination with the mechanisms of learning. This led him to make a robot that could be taught like Pavlov's dog and learn to associate a whistle or a kick with a specific event.

In 1935, however, all this lay in the future. Walter was able to stay on at Cambridge and continue his work because he won a Rockefeller fellowship that funded his work on physiology and electrical brain activity. He travelled to Germany to visit the laboratory of Hans Berger, the man who discovered that there were regular waves of electrical activity in the brain that could be measured. Walter refined Berger's work, developing a number of devices that helped measure these brain waves with more accuracy.

While he got on with his studies Walter was not neglecting his social life. It was at Cambridge that Walter met people that shared his outlook and his character began to develop. As a postgraduate Walter was recruited into the Apostles, the most secretive of Cambridge's secret societies. Walter came to it late, many of its members were recruited while undergraduates, but he arrived in style. The two men who sponsored his membership were the leading lights of the club at that time. They were Victor Rothschild and Guy Burgess. One came from a rich family and the other was an aesthete who was already dedicated to the communist cause. They were fairly typical members of the Apostles. One being rich and well connected, the other brilliant and gay.

Those who were asked to join the Apostles usually had some quality that made them stand out, they were either brilliant, beautiful or bolshy, sometimes all three. Walter was probably picked because he was bright and not afraid to speak his mind or flout convention. The fact that his father had links with British Intelligence and European anarchists probably did him no harm either.

The Apostles was primarily an essay club. Every week a member would read a paper they had prepared, which would then be debated. The issues discussed were secret but there is no mystery about the fact that most of the members of the Apostles held opinions that were not shared by society at large in the late 1930s. Many of its members were gay, communists, or both. Membership of neither group could be openly admitted.

It was not just that the Apostles held unfashionable opinions – how they reached these views and why they believed them was just as important as the opinions themselves. The Apostles saw themselves as pursuers of truth but not just one truth. They recognized that people believe different things and that their reasons for doing so may be just as valid as the reasons we have for following our own. The Apostles were interested in finding which opinions were well founded and worthy of belief rather than hollow products of fashion or taste. One historian of the society wrote of their outlook: 'Since the whole force of conventional and sectarian thinking is powerfully opposed to such thoughtful, patient exploration of why people think what they do, the Apostles felt that they were freeing themselves from the restraints of ordinary social relations.'[23] J.M. Kemble, a former member, says: 'From the Apostles I, at least, first learned to think as a free man.'[24]

Part of the reason that clarity of thought was so important at this time was because of the great changes that seemed to be taking place in society. The decade had opened with the Wall Street Crash and since then things only seemed to be getting worse. Unemployment was on the increase, society was coming apart at the seams and although the times seemed to demand fresh thinking governments seemed paralysed and unable to make a difference.

The prevailing feeling was summed up in 1932 when John Strachy published a book called *The Coming Struggle for Power*. In it he wrote: 'The capitalist system is dying and cannot be revived. Religion, literature, art, science, the whole of the human heritage of knowledge will be transformed. And the new forms, whether higher or lower, which these principal concepts of man's imagination will assume, will depend on what new economic system will succeed the capitalist system.'[25]

When turning away we can go either to the left or right. So it was in the 1930s. Many saw capitalism crumbling around them and turned to communism or fascism as the only viable alternatives. Countries too seemed to be making the same choices, as the rise of fascist governments showed. In Britain the *Daily Mail* came out in support of Italian and German fascism and for years the *Daily Express* beat a nationalist drum.

The Apostles had already made their choice. Anthony Blunt says it was not a difficult one to make in the 1930s. Later he wrote that Russia and the Communist Party were the only ones working against fascism since Western governments did not seem to know what to do about the rising nationalist threat. 'I was persuaded . . . that I could best serve the cause of anti-fascism by [working] for the Russians.'[26]

Kim Philby was the same. He wrote in his memoirs: 'It cannot be so surprising that I adopted the Communist viewpoint in the thirties; so many of my contemporaries made the same choice.'[27]

Walter was certainly a sympathizer to the communist cause. His son, Nicolas Walter, describes him as a 'fellow traveller', someone who was sympathetic but never became a member of the Communist Party. He was never an ideologue like Blunt or Burgess. Walter did keep dangerous company though and stayed a member of the Apostles all his life, attending the yearly reunions as long as he was able.

Because he mixed with spies and communists Walter did not escape the attentions of the security forces. Their interest reached a peak when Guy Burgess and Donald Maclean defected in 1951 and again in late 1963 when Anthony Blunt was unmasked as a Soviet spy. Blunt was the surveyor of the Queen's pictures and an international art historian and his work as a spy for Russia only became public knowledge in 1979. But, as Peter Wright, former assistant director of MI5, makes clear in his autobiography, *Spy-*

catcher, the secret services had known about Blunt long before this.[28] Walter was not interrogated, in fact Nicolas Walter says the security forces treated him with uncommon courtesy. They carried out all their questioning over lunch at the Army and Navy Club in Whitehall.

It was unlikely that he gave anything away, the loyalty of the Apostles was legendary even before the days of Burgess, Philby and Blunt. When journalist Tom Driberg questioned Guy Burgess in Moscow in the 1950s about the Apostles, Burgess, although he had betrayed his country, could not bring himself to do the same to the Apostles.[29]

Walter had no secrets to tell about his friends nor was he ever a spy, even though he had all the right qualities for a secret agent. He was fluent in three languages, a keen glider pilot and scuba diver. He was charming and intelligent, he even had colleagues in the Soviet Union. Perhaps the only reason he was not a spy was because he had no secrets and was never in a position to acquire them.

One thing Walter did share with his communist friends was their faith that science and reason could save mankind from ruin. The USSR seemed to be one of the few states that took the intelligentsia seriously. Nicolas Walter says that his father admired the USSR because, 'It took scientists very seriously, gave them high salaries and privileges.' But this was as far as his commitment went. Later in life Walter got to know some Soviet scientists well and visited the USSR several times. He was never tempted to live there though, largely because he would have been stifled by communism.

Ideal Home

Science in the service of man was the main theme of Walter's science-fiction novel *Further Outlook*. The book is a homage to Olaf Stapledon, whom Walter calls 'the Virgil of science-fiction',[30] and who wrote several novels with grand sweeping themes and huge timescales.

Taken at face value *Further Outlook* is a fairly run-of-the-mill treatment of a very old plot. The story revolves around the discovery of a strange craft that turns out to be from the future and flown here by the very man who found it. Read more deeply, the book is a good guide to Walter's outlook on life. The narrator of the book is a psychiatrist called Dr Wing Wedge, a thinly disguised William Walter.

The four central characters in the book, three men and a woman, have what would now be called an open relationship. In real life Walter was married at least twice. He never divorced his second wife, Vivian Dovey, but left her in 1960 to live with a woman called Lorraine Aldridge. To make it look like they were married Aldridge changed her name by deed poll. In fact she was still married to another man.

In the novel Wing Wedge declares his principles: 'Any man or woman who is a scientist by avocation as well as by education is a natural anarchist; the laws of science are not man-made laws; they are the parameters of man's hypothetical vision of the orderly processes of life and expendable in the first moment of inadequacy or redundancy. The scientist is always wrong; he is essentially a rebel; he only postulates his "laws" for them to be broken by himself or by others.'[31] Certainly Walter flouted convention and liked to see himself as something of an anarchist. He liked to provoke people and dressed down when he was expected to dress up and happily insulted people he had no time for. Although he liked to think of himself as an intellectual anarchist Nicolas Walter says he was less an anarchist and more an intellectual snob who simply wanted the freedom to go his own way.

Burden of Proof

In 1939 Walter got his chance to study more of what interested him when he took up a post as director of physiology at the newly opened Burden Neurological Institute (BNI) in Bristol. He was to work at the BNI for the rest of his career.

It was there that he made many of the major discoveries of his brilliant career in electro-encephalography. He was the first to measure delta rhythms, the slow electrical activity that surrounds brain tumours; he found that the brain waves of epileptics undergoing seizures were abnormal; and he discovered the theta rhythms that seem to be associated with mood and imagination. The Burden was also a pioneer in the use of electro-convulsive therapy. It was the first place in Britain that the technique was tried.

It was also at the BNI that Walter built his pioneering robots. During the war years Cambridge psychologist Kenneth Craik visited the BNI to see if he could use the automatic analysers Walter used to crank through the reams of brain wave data he regularly collected. Craik wanted to use the machines to analyse data drawn from the aiming errors of air gunners.[32] The talk with Craik dwelt on goal-seeking missiles and scanning mechanisms – whether to spot the missiles or trends in data Walter does not say. What he does say is that by combining the two ideas he thought he might have the makings of a simple animal.[33]

Walter also had ideas about how the nervous system of such a simple creature should be constructed based on his work with nerve physiology. He was convinced that the richness of human behaviour and experience was due to the number of connections between neurons rather than their sheer numbers. He proposed to give his electromechanical animal the smallest possible number of neural elements (two) and see what kind of complex behaviour emerged. He says: 'The number of components in the device was deliberately restricted to two in order to discover what degree of complexity of behaviour and independence could be achieved with the smallest number of elements . . .'[34]

The first artificial animals he built were called Elmer and Elsie. He gave the species of robot a mock Latin name, *Machina speculatrix*, because 'it explores its environment actively, persistently, systematically as most animals do'.[35] The animals were also known as 'tortoises' because their electromechanical innards were shrouded by a clear plastic shell.

M. speculatrix was deceptively simple. It had only two sensory receptors and two effectors[36] connected by two artificial nerves. The sensors were a light-sensitive cell and a contact switch that sat between the shell and the chassis of the tortoise. The two effectors were the front driving wheel and the steering motor.

The photocell was attached to the steering column, which spun constantly when the robot was in the dark. This acted as a scanning mechanism to search for light sources. When light entered the cell scanning stopped and the drive motor kicked in to move the robot towards the light. When bright light fell on the photocell the scanning mechanism started again at half speed. This made the robot slowly approach the light source.

The contact switch helped the robot cope with obstacles and steep slopes. When the shell bumped up against a block or rock in its path the switch was closed and re-routed some of the internal connections of the robot. This made the robot stop moving forward towards a light source and produced a turn and push manoeuvre. Minor obstacles were gently pushed out of the way. Heavy objects that the robot could not nudge out of the way were gradually worked around. Once it was clear of an object the robot would once again start scanning for light sources.

Despite the simplicity of its construction M. *speculatrix* exhibited a wide variety of behaviours. It explored the ground in a series of swooping curves and was attracted to lights that were not too bright. If dazzled, it turned away. But all these behaviours were built in and expected. The most interesting things emerged when M. *speculatrix* had a small light of its own placed on its front. This light was extinguished when the robot was heading for another light source but when it was just exploring the bulb was turned on.

When M. *speculatrix* was placed in front of a mirror the reflected light hit the photocell so the robot stopped looking and headed for the mirror. But this action turned off the light so M. *speculatrix* started searching again. The result was astonishing. 'The creature therefore lingers before a mirror, flickering, twittering, and jigging like a clumsy Narcissus. The behaviour of a creature thus engaged with its own reflection is quite specific, and on a purely empirical basis, if it were observed in an animal, might be accepted as evidence of some degree of self-awareness. In this way the machine is superior to many quite 'high' animals who usually treat their reflection as if it were another animal, if they accept it at all.'[37]

When two M. *speculatrix* were left to interact equally interesting things happened. Both the robots were attracted to each other's light. But both turned off their lights when they turned towards the other robot so the two never met. They chased each other around momentarily before losing interest. If there was no other light for them to react to this mutual dance continued until the batteries of both robots were exhausted.

The behaviour of the robots was certainly lifelike, so much so that they caused some consternation in the Walter household. He says M. *speculatrix* behaved 'so much like an animal that it has been known to drive a not usually timid lady upstairs to lock herself in her bedroom, an interesting blend of magic and science.'[38] Only now are more modern robot makers getting to the same point.[39]

Although the analysis of these behaviours was written forty years ago it shows that Walter was interested in exactly the same topics as current ALife researchers. M. *speculatrix* has only two elements, the smallest interesting number, but from their interconnection emerges unexpected behaviours. But Walter makes the point that part of the complexity of the behaviour is due to the environment that the robot finds itself in and the willingness of the observer to impute intelligence to the robot. Today this is seen as one of the key insights of ALife.

Walter also made other robots. One he called M. *docilis* to signify that it was teachable and could learn to associate sounds and touches with events. He created a circuit that could become a conditioned reflex. *M. docilis* was a step up from *M. speculatrix* and Walter taught it lots of simple tricks. The robot was taught to come forward when a whistle was blown. It was conditioned to move forward because previously every time the whistle was blown a light was shone on its photoreceptor. The whistle had to be blown at the same time the light was shown or the reflex took far longer to establish. Walter soon tired of teaching such a simple trick and he moved on to blowing the whistle and kicking the shell of the robot a few times. Soon *M. docilis* learned that a whistle blast meant an obstacle was in the way and it backed away before the blow fell. 'After five or six beatings, whenever the whistle was blown the model turned and backed from an "imagined" obstacle.'[40]

Walter only built two species of the tortoises but it was enough. The robots were an instant hit and were written up in the national press. In 1949 the *Daily Express* speculated that they would soon be finding their way into Christmas stockings.

Partly because of the success of *The Living Brain* and his work with Elmer and Elsie, Walter became something of a celebrity in 1950s Britain. He regularly appeared on the Brains Trust – a scientific debate programme that, his son reports, he always enjoyed. In 1961 he was even on Desert Island Discs. Nicolas Walter says that being invited on this programme caused something of a crisis because Grey Walter could not think of six pieces of music he liked. He was a dedicated scientist with little interest in anything outside his field. He asked all his friends to think of pieces for him and he talked about these when he appeared on the programme.

But his popularity was short-lived and the robots were never made into toys. They languished largely forgotten until 1996 when Owen Holland from the University of the West of England rediscovered one of them in the basement of Nicolas Walter's house. Since then Holland has worked to restore the robots and 'make Grey Walter's work more available to the scientific community in order to enable it to be assessed and appreciated in a way which does justice to the man and his ideas.'[41]

Walter wanted to go further with his robots and try and give them even more behaviours. He wanted to use transistors rather than valves to make much more elaborate creatures. He wrote: 'There seems to be no limit to which this miniaturization could go. Already designers are thinking in terms of circuits in which the actual scale of the active elements will not be much larger, perhaps even smaller, than the nerve cells of the living brain itself. This opens a truly fantastic vista of exploration and high adventure . . .'[42]

When he built these robots Walter was years ahead of his time. In 1984 Valentino Braitenberg published a book called *Vehicles*, which concerned a whole family of imaginary robots that would appear to act purposefully despite being of very simple construction. Walter built robots that did what Braitenberg only imagined, and he did it in the 1950s. Even today Walter's robots invite respect. Rodney Brooks, one of the leading robot makers in the world today, says of them: 'The complexities and abilities of Walter's physically embodied machines rank with the purely imaginary ones in the first half dozen chapters of Braitenberg three decades later.'[43]

When talking about the behaviour he observed Walter used ideas and language that would not be out of place at modern-day ALife conferences. What stopped Walter going further was the limitations of 1950s technology as much as a lack of useful theories to underpin the direction he wanted to follow. At the time Walter was working the field of AI was only just getting under way and despite his success researchers turned away from the difficulties of building robots to the easier to manipulate world of computers and software. Because of this Walter's work remained largely forgotten until the ALife movement rediscovered it.

Sadly Walter did not live to see his work gain the recognition it has today. In 1970 he was on his scooter when he collided with a runaway horse. He survived the accident but the horrendous head injuries he suffered in the crash left him a shadow of his former self. He lost his sense of smell and the sight in one eye and the injuries wrought great changes in his personality. To everyone who knew him he was a different person after the crash. Ray Cooper, who had been Walter's collaborator throughout his time at the BNI, wrote, 'The spark that had burned bright for forty years had gone out.'[44]

Walter lived on seven more years after the crash but made no more contributions to science. He died suddenly of a heart attack in 1977.

You Started It

While Walter's spark may have dimmed, his work has inspired others to study ALife who in their turn have made major contributions to the field. One man who would not have got involved in ALife if it was not for his early encounter with Walter's work is robot maker Rodney Brooks.

Brooks was born in Adelaide in 1954, the year after Grey Walter's book *The Living Brain* was first printed. It was a book that Brooks was to become very familiar with. He even went as far as to build his own versions of Walter's tortoises. Rather than copy them exactly though, he replaced the relays that Walter had used with transistors. His childhood tinkering did not stop with the building of replica robots. At the age of ten he built his first computer, a tic-tac-toe machine that was made out of ice cream tins, nails, light bulbs and batteries.[45]

Unfortunately in Australia in the late 1960s and 1970s there were no places that Brooks could go to learn more about robotics. Instead he embarked on an orthodox science degree and started studying mathematics at Flinders University on the outskirts of Adelaide in Southern Australia. For these few years his ambition of building machines remained unfulfilled. He got back on track once he won a fellowship to study in the US. He secured a post as an assistant at Stanford University in California where he shared an office with another keen robot maker, Hans Moravec.[46]

Among maverick robot makers and ALife researchers Moravec is in a class by himself. He often unashamedly talks up the prospects of ALife research. In 1987 at the first ALife conference he calmly predicted that in fifty years the field would produce an artificial human. The first line of his book *Mind Children* makes the same prophecy. The book is an exploration of what Moravec calls the post-biological world and the marvels we will make once we have freed ourselves from the tyranny of our genes. It is both a catalogue of what Moravec believes we will be creating in the next fifty years and an exploration of current-day robotics. Alongside this are short fantasies written from the perspective of those living fifty years hence.

Moravec started studying robotics at Stanford in the

1970s and in 1996 he was contracted by NASA to work on the design of a futuristic robot that would be useful on manned, deep-space missions. He based his ideas on the robotic bush that makes an appearance in the science-fiction novel *Flight of the Dragonfly* by Robert Forward.

Back then Moravec was working on making a robot carry out a more mundane task – crossing a room without bumping into any of the chairs, rubbish bins and people that might get in its way. This was exactly what Brooks wanted to do, but he doubted that Moravec was going about it in the right way. Moravec's robots often took hours to get from one wall to another. What took the robot so long was cranking through the computer program it used to work out where it should go next. It spent more time just sitting thinking than it did moving.[47]

For their time Moravec's robots were leading edge, but Brooks was unimpressed. His disillusionment only grew as he moved on to other jobs at robot labs around the US. He bounced between Stanford, Carnegie-Mellon and MIT until 1984, when he finally settled in Massachusetts as an assistant professor at MIT's AI lab. With the job went an obligation to find the funding and researchers for a mobile robot laboratory. Brooks was only too happy to take on the challenge.[48]

Brooks soon found workers for the lab, among them ex-US Naval Academy graduate Anita Flynn and graduate student Jonathan Connell, and funding to build a robot. They even had a name for the robot. They decided to call it Allen after Allen Newell – one of the pioneers of AI. What Brooks lacked though was a research direction. He knew he didn't want to build robots as slow and stupid as those that others were working on, but he did not know what he had to do differently.

At the time Brooks was married to a Thai woman and shortly after the Mobot lab was set up, he and his wife went to visit her relatives in the rural south of Thailand. None of his in-laws spoke English so he could not talk to them. He couldn't amuse himself exploring the countryside either because he was warned that it would be unwise to go exploring by himself. So for a month he just sat around and thought about robots in general and Allen in particular. He reflected on the slow, stupid and stumbling robots he had seen and their failings. At the end of the month he had his research programme and an accompanying philosophy. He realized that everything established AI researchers were doing was wrong. In Brooks' opinion the Good Old Fashioned AI (GOFAI) philosophy, experimental approach and conception of how the brain works was never going to produce machines that could be called intelligent.[49] To understand why he came to this conclusion we have to look into the history of AI.

Searching for Answers

Before engineers, computer programmers and technologists hijacked AI, it was dominated by philosophers. For them it was known as the mind–body problem. The problem was, and still is, that no one could work out how the physical parts that make up our body could give rise to something as insubstantial and spiritual as thoughts. Even if, as some reasoned, our thoughts were something non-physical such as the soul, there still remained the problem of explaining how such soul-stuff could interact with the physical body. If one was physical and the other not, how could they affect one another?

For a long time research into the qualitative differences between the solid, fleshy body and the spiritual mind involved a philosopher sitting in an armchair and thinking really hard. They did this because they assumed that we have a uniquely privileged access to our thought processes. Because of this they thought that the best way to study the workings of the mind was to do a lot of thinking and see how it felt. This attitude still prevails among many philosophers and psychologists.[50] These scientists find it hard to believe that we can be deceived about the way that our minds work. But there is a growing body of evidence that shows we are deluded about what we see at all times and that we often know nothing about the way our brains really work.

Back in the seventeenth century the thought that humans could be deceived by their own brains was heretical and thinking seemed an appropriate method of research. One of the best explorations of the mind–body problem came about this way. The *Meditations* of René Descartes were published in 1641 and are written from the perspective of a man, Descartes, sitting and thinking about what he can believe, what he can be sure about. He eventually concludes that he can be sure of nothing but the fact that he is thinking and therefore that he exists or *cogito ergo sum*, as Descartes' famous phrase has it.

For himself, Descartes was convinced that the soul or mind was the one thing that set man apart from the animals. All those living things without a soul, or the pineal gland through which it was supposed to operate, he regarded as mere machines, on a par with clocks. For him the screaming and yowling of an injured dog, cat or cow was akin to the grinding of gears in an engine.

Although Descartes' *Meditations* is a classic treatment of

the mind–body problem and a typical example of how the problem was tackled, Descartes was one of the few philosophers who, apparently, troubled himself to build a mechanical replica of a human being, presumably in order to demonstrate the difference between humans and machines. The automaton was called Francine, named after his daughter, who died in 1640 when she was only five years old. This early robot barely lasted that long. On the voyage to show off his mechanical marvel in Holland, the captain of the ship looked through Descartes' luggage and was so shocked and scared by the lifelike appearance of Francine that he had the thing thrown overboard.[51]

In the three centuries between Descartes' *Meditations* and the Dartmouth conference opinions about bodies and minds changed as our knowledge of the workings of the brain grew. For a start we now know that the pineal gland is involved in the regulation of sleep cycles. But one thing has remained the same. With very few exceptions everyone studying minds and brains has begun with the human brain and worked from the top down.

Room at the Top

There were good reasons for doing this. For seventeenth-century philosophers like Descartes it made sense because humans were the only creatures with souls and therefore the only ones with a mind–body problem. In the twentieth century interest still centred on human brains because the problem remains, albeit in a different form. Few people are now looking for ways to explain how the soul can interact with the brain, now it is a quest to 'explain cleverness . . . , in terms of suitable orchestrated throngs of stupid things'.[52] The human cortex is much deeper and has far more folds than that of any other animal. By beginning at the top and working down it was thought that it would be easier to expose and extract the essence of human thinking and then replicate it elsewhere.

Early AI researchers were not naïve enough to study the brain as one single entity. The billions of neurons it is made of are daunting enough today and were much more so in the 1940s and 50s.

The technology of the time was not up to making a robot see or hear, so early AI researchers concerned themselves with abstract realms of thought such as chess playing or geometry. They did this partly because an ability to play chess well was seen as a considerable intellectual challenge and partly because they thought that success was likely to come quicker if they did not have to engage the world when building thinking machines. Recognizing the scale of the task they were setting themselves they decided to tackle it in piecemeal fashion. This fitted with their conception of how the brain worked. They conceived different functions such as language and vision to be situated in separate areas of the brain. Because of this, trying to replicate thought processes in isolation did not seem too great a crime. To make it even easier they decided to tackle those aspects of intelligence and thought that were entirely internal and needed no connection with the outside world.

The Dartmouth conference of 1956 was important for many reasons. Those attending the seminars, men such as Marvin Minsky, John McCarthy, Herbert Simon and Allen Newell, were the leading thinkers in the field and they emerged from the weeks of debate with a very definite view about the brain and the best way to study it. Many future AI researchers had their postgraduate studies supervised by Dartmouth men so their views have been sustained for years.

For this influential group the brain was primarily a really good searching mechanism. They thought that human cognition, in the abstract realms they decided to study, was a matter of searching through all the possible explanations for an event and choosing the one that fitted best or made the most sense in the circumstances. With chess this would be a move that took the player making it closer to victory, in geometry closer to producing a proof. The history of AI for the thirty or so years following Dartmouth can be summed up as a quest to find out how humans choose between possible explanations for an event. If it proved impossible to emulate human search mechanisms then these researchers had a second goal, which was to create methods of scanning through a huge list of alternatives that approached human-like levels.

Often this searching for answers involved comparing what the computer was being told with some internal memory store made up of symbols. Once the computer knew what it was dealing with it would work out a plan of action and then carry it out. Then it would take more input from its sensors, to see how the world had changed or how its chess opponent responded, and the cycle would begin again.

The early AI researchers had good reasons for doing this. They thought the brain worked in the same way. It was widely believed that inside the head of every person was a miniature model of the world. In this model world everything was represented symbolically. Memory was believed to be made up of a huge collection of symbols, each one representing something, anything and everything that we encounter during our daily lives. They reasoned that we manipulate the symbols and the model when deciding what to do. Once we have proved the plan works in our heads we pass the plan to our limbs and carry it out.

Hand in hand with this belief in the importance of search mechanisms went a faith in computer power. It is an accident of history that at the same time as AI was getting off the ground computers were starting to be used seriously in academia and business. The power of computers was progressing in leaps and bounds.

The belief was that as computers became more powerful it would become easier to mimic the brain. When a suitably powerful computer was coupled with a fast search program an intelligent machine would be the result. Herbert Simon wrote in 1957: 'It is not my aim to surprise or shock you – but the simplest way I can summarize is to say that there are now in the world machines that think, that learn and that create. Moreover, their ability to do these things is going to increase rapidly until – in a visible future – the range of problems they can handle will be coextensive with the range to which the human mind has been applied.'[53]

Early successes with programs such as Herbert Simon's General Problem Solver, Feigenbaum's Elementary Perceiver and Memorizer and Gelertner's Geometry Theorem Prover, convinced the AI community that it was on the right track. They were even tempted to start experimenting with robots such as Shakey, but this is where success became harder to find. The only notable recent success this approach has had took place in 1997 when IBM's Deep Blue computer managed to beat world champion Garry Kasparov at chess. This computer plays chess very differently to a human. It relies on thirty-two high-speed processors operating simultaneously, each one coordinating sixteen special-purpose 'chess chips', to search through 200 million positions a second and work out possible ways the game could develop given the past moves. All these possible moves are given a score and Deep Blue then picks the move that takes it closer to victory. Humans usually only work a few moves ahead, looking for patterns and applying rules. Deep Blue is just good at searching.

It has been said that what threw Kasparov off was that he could not indulge in some of the intimidating tactics that he employs when playing other humans and that a mistake he made in an early game gave Deep Blue the edge. Perhaps, but as Kasparov says this should not be taken as an excuse. He accepted the conditions. When writing about the experience of playing Deep Blue Kasparov said that he 'sensed a new kind of intelligence' was operating inside the computer. It wasn't human, just different. 'Although I did see some signs of intelligence, it's a weird kind, an inefficient, inflexible kind that makes me think that I have a few years left yet.'[54] This was written before he lost to Deep Blue.

Despite the success of Deep Blue, Brooks believes that this concentration on search methods and fast computers is where the field has gone astray. For him all AI has been doing since the 1956 Dartmouth conference is look for ways to improve searching through the growing number of possibilities. John McCarthy wrote the LISP computer language to make it easier to write computer programs that supported large searches. Many of the robots built at this time, such as Shakey and Freddy, were placed inside simplified worlds because this cut down the number of interpretations a robot could give to what it was seeing.

Part of the reason that search systems of the time were so

slow was because of the serial way in which they processed information. Computers may have been getting increasingly powerful but they did their work in the same way, one piece at a time. Once they had done as much as they could with one piece of data they moved on to the next part. There was none of the massive parallelism that is found in living systems. In the human brain many different processing centres work on a problem at the same time.

Brooks believes that AI researchers forgot why exactly they were studying chess playing rather than walking or perception. It was because years earlier it had looked like it was the most straightforward of mental tasks to tackle, not because it was the essence of mental life.

At the same time the field has been working with a promissory note for computer power. It seems as if there is never quite enough computer power to do the search fast enough and produce intelligent reactions, but there was enough of an increase in computer power to reassure AI researchers that they were getting somewhere. As Brooks has written: 'performance increases in computers were able to feed researchers with a steadily larger search space, enabling them to feel that they were making progress as the years went by.'[55] For Brooks those years were largely wasted. They only served to show how not to study AI.

Part of the reason that the progress of robot makers has been slow is because they started off by creating a sheltered, simplified world and then tried to take what had worked there into the buzzing confusion of the real world. But robots that excelled in the model world failed miserably when let loose in the wild.

Brooks has taken a different approach. As he himself has put it, 'Our ideas took a route different from the traditional approach in Artificial Intelligence.'[56] The remark is uncharacteristically kind to AI. At other times Brooks has dismissed what it has achieved as a waste of time. In a recent paper he has written that for the last fifteen years '[AI] has been thundering along on inertia more than anything else'.[57] The MIT mobile robot workers threw out the notion of planning, central representation and impractical models of the world. They do not cheat with simplified environments.

Brooks is being rather too unkind to GOFAI because it has undoubtedly produced many useful devices and programs. Now expert systems are widely used to help do everything from make computers to help diagnose heart disease. Even the crude robots that came out of GOFAI blazed the trail for the use of robots on production lines. The same narrow, precise movements are also starting to be used in robots that will be able to perform surgery. The great advantage these robots will have is a steady hand at all times and greater precision than any human surgeon can achieve.

But Brooks is unrepentant. He goes as far as to say that many of the people working in AI today are trying to solve problems that have nothing to do with intelligence or intelligent behaviour.[58] Even if these problems are solved he doubts they will bring us any closer to understanding what intelligence is, how it emerges and how to emulate it. The best they will do is remove some experimental obstacles that have been holding back a few AI researchers.

It was not always so. Brooks points out that during the early years of AI there were signs that people were willing to tackle real world issues. When Alan Turing, Norbert Wiener, John von Neumann, Ross Ashby, Claude Shannon and Grey Walter considered what it would take to produce an intelligent robot they all assumed that it would need some way to interact directly with the world. Following this line Walter made his tortoises and Ashby made his own self-regulating robot called homeostat. They found it hard to pursue this avenue of research because the technology of the time made the creation of such a robot too formidable a task. Hand in hand with this were some conceptual issues that made it difficult to go beyond anything other than the simplest of robots. Even so Brooks says the early work of Walter in particular was thirty years ahead of its time.[59]

This early work was ignored by the larger AI community. They turned inwards and away from robots that dealt directly with the world. But this is where Brooks and his co-workers begin, in the real world, in the wild, with nature.

Fast Elephants and Earwigs

What sets Brooks apart from the GOFAI community is his dogged insistence on dealing with the real world. For him this 'situatedness' is key, no simulations will do. This way 'there is no room for cheating'.[60] Brooks has good reasons for insisting on this point.

Firstly, simulations shield researchers from uncomfortable realities. Problems that were trivial in a computer simulation or a simplified world can become crippling once the robot is built and let loose. Having a robot succeed in solving the same problems that animals face is confirmation that the approach it represents is valid and that the thing works.

Secondly, there is a real advantage to building a robot that walks or drives around the lab because this removes the need for any internal model building. The robot simply responds to what it finds rather than tries to remember what the world looks like. The world can simply be taken for what it is because it 'really is a rather good model of itself'.[61] This concept of using the world as your memory bank is used widely in the animal kingdom, ants in particular exploit the world in this way (see Chapter Five).

Brooks' interest in the real world does not end with having his robots wander around up to their metal knees in it. To get his robots started Brooks has studied insects and is happy to steal ideas from nature. He sees little point reinventing the leg or feeler.

Hand in hand with the emphasis on situatedness and being in the world is a similar insistence on embodiment and autonomy. By this Brooks means that the robots must have a body and be able to wander around the world at will. There should be no umbilical cords to computers that do all the robots' thinking for them. This is important because the real world is where the meanings we share are grounded. Connecting with and acting on the world gives meaning to what a robot does. The physical world is where our mental abstractions 'bottom out'. Everything animals do has to take account of the facts of existence. They, and we, learn from the physical world what it means to be hot, hungry or frustrated. A brick in the path of a robot finding its way across a room cannot be wished away, it demands attention. Before humans ever enter a school building they will have spent years being educated by the world. In exactly the same way, when a robot learns to cope with stubborn objects, the interaction gives meaning to what it does. 'Without an ongoing participation and perception of the world there is no meaning for an agent. Everything is random symbols.'[62] It also means we have to consider putting robots in with the living rather than the dead. They have crossed the divide. They are in this with us.

Giving a robot a body also means that the connection between world and agent is concrete. None of Brooks' mobile robots can avoid the world. At the most fundamental level they are connected to it. In every mobile robot built by the MIT workers the lowest level of control connects sensors that are looking out at the world to the systems that control the legs or wheels the robot uses to get around. This helps the robot avoid bricks, people and furniture.

One of the first robots built by the Mobot lab to demonstrate the new approach was called Allen. This waist-high robot, described memorably as 'R2-D2 in its electronic underwear',[63] sat happily in the centre of a room until someone or something approached it. Then it would scurry away, neatly avoiding obstacles as it made its getaway. To give it the ability to do this Brooks fitted Allen with sonar and sensors to detect the returning signals. By picking up the signals and comparing how long they took to return, Allen could find a clear escape route. The robot knew when a static object was looming because the signals it was bouncing off it would arrive quicker. The sonar sensor was connected directly to the robot's wheels so evasive action could be taken quickly.

Once a robot had been given these low-level abilities, Brooks and his colleagues started working on other layers of behaviour. In stark contrast to the GOFAI approach, where the outputs of one level of control become the inputs to another, Brooks isolated every level. There was no interaction between them. Once something was proved to work it was left alone and never tinkered with again. Brooks had a good reason for doing this, the same thing happens in evolution. Once an animal has acquired an adaptation it is rarely discarded. The same is true of one of Brooks' mobile robots.

This is not to say that the outputs from separate behaviours never conflict, they do. The outputs from all the sensors and higher level behaviours are brought together into a common short-term memory store. As the results of one sensing-processing loop arrives it overwrites any data that was there already. Once the robot has finished doing what was taking up its attention it moves on to the next instructions it finds in this memory store. Computationally the robot is behaving like a finite state machine. As it carries out different behaviours it is kicked into a separate state. The rules combine to produce novelty, innovation and surprise, just like in a cellular automaton, and real life.

One of the key ideas in ALife is the recognition that the essence of any machine or computational system can be abstracted. What a machine does can be isolated from the physical device and the way it does it, but nothing will be lost in the process. This is the key point to take away from Turing's work. Consequently these abstract specifications can be transferred to any number of other formally equivalent systems. If we accept that living things are collections of physical finite state machines then we have to grant that there are properties of them that can be abstracted and captured in other systems, such as robots. As Chris Langton puts it: 'the principle assumption made in Artificial Life is that the "logical form" of an organism can be separated from its material basis of construction, and that 'aliveness' will be found to be a property of the former, not of the latter.'[64]

Brooks calls his method of building up behaviours in

layers the subsumption architecture because every new layer subsumes control to another layer when conditions demand. He reasoned that out of enough layers would emerge intelligent action. All this with no internal representation.

Using this approach the MIT mobile robot group has managed to create robots that exhibit some very complex behaviour. They made robots that were smarter than any seen before and they did it in weeks. Shakey, Ambler and their ilk took far longer to develop yet were not as intelligent as any of the fast, cheap and out-of-control robots running round the MIT labs.

One of the first robots produced at Brooks' Mobile Robot lab performed a vital function in a workshop full of graduate students, hackers and technologists: it collected up empty drink cans. Herbert (named after AI pioneer Herbert Simon) was built by graduate student Jonathan Connell and is typical of the 100 or so robots that have walked and rolled out of the Mobot lab. Herbert was equipped with thirty infra-red sensors to help it avoid obstacles, it had a laser-based vision system for object recognition and an arm with a hand on the end to pick up cans. Using information from its IR sensors Herbert would move around the lab, following walls, avoiding obstacles and moving through doors as it looked for objects of interest.

The beam of the 7mW helium-neon laser was fed through a cylindrical lens to spread it into a flat plane. This beam was bounced off a moving mirror and the area being scanned ranged from 10 degrees above the horizontal to 48 degrees below to give the robot a relatively deep visual field. The field was scanned many times a second to give the robot a good idea of what it was approaching. Once it saw something of interest it would approach cautiously making sure the object was what it thought it was: hopefully a tin can. When the robot got close enough it stopped moving. Then the arm would start extending.

Although the hand was physically connected to the arm, it was not activated by it. The hand was autonomous and closes when something breaks an infra-red beam passing between its open fingers. As soon as the hand closes the arm retracts and the robot moves off back to where it came from, dumps the can and starts searching again.

In basing its robots on real animals the Mobot lab has clearly started a movement of sorts. In early 1998 the finishing touches were being put to an ambitious artificial animal by John Kumph – a graduate student in MIT's Ocean Engineering department. Following on from work with a tethered robot tuna, Kumph was creating a free-swimming robot pike. He picked pike because as predators they are renowned for their speed and agility in the water when ambushing and pursuing their prey. Making a robot perform anything like as well as a real pike would be quite an achievement.

He picked a pike species known as the chain pickerel for no other reason than it is one of the smaller pike species and would fit well into the 80,000-gallon tank owned by the Ocean Engineering lab. A couple of years work has produced a 7.5 kg, 80 centimetre pike that swims and turns well and is fully autonomous. A school of the robot pike or other fish would be useful to marine biologists studying migrating fish or the ocean depths.

Another reason for studying fish, especially robotic ones, is that they represent something of a paradox because they can swim faster than they should. Studies of fish have revealed that they do not seem to have the energy to swim at the speeds they reach nor to travel the distances they do given their muscle power. Kumph is studying the pike to see whether this is because of the way a fish swims or whether it somehow smooths the flow of water over its scales to avoid creating too much drag.

Genghis was the name given to another robot that, until 1997, was the star of the lab. It was called Genghis 'because it stomps over things'. This 1 kg, six-legged robot was one of the most sophisticated built at MIT and it uses legs rather than wheels to get around. Although the legs work together they are controlled independently, there is no central system that coordinates their movement. The legs are given a few simple behaviours so they know what to do in the different situations they find themselves in, such as when they are raised, down or trailing. Again this mimics the real world. As knowledge about the brain and nervous system has grown it has become obvious that intelligence is not concentrated in clumps, it is spread throughout the system.

One of the most basic behaviours that each leg possesses can be characterized as follows: 'If I'm a leg and I'm up, put myself down.' There are other behaviours such as 'If I'm a leg and I'm forward, put the other five legs back a little' as well as 'If I'm a leg and I'm up then swing forward'.[65] No central system controls the order in which the legs carry out these behaviours.

'To create walking then, there just needs to be a sequencing of lifting legs. As soon as a leg is raised it automatically swings forward, and also down. But the act of swinging forward causes all the other legs to move back a little. Since those legs happen to be touching the ground, the body moves forward. Now the leg notices that it is up in the air, and so puts itself down. The process repeats by lifting the next leg and so on, and the robot starts to walk.'[66]

Genghis is not just an out-of-control stumbler. Once the basic walking behaviour had been created Brooks, Connell and colleagues fitted the robot with more sensors. Some of these monitor the pitch and roll of the robot's body – the degree to which it sways from side to side and along its length. Other sensors track the forces acting on the legs as it pushes itself along. Using the information from these sensors other behaviours have been created that help Genghis walk better. No modifications are made to the original walking behaviour. The new behaviours simply override the basic walking behaviour as conditions demand. When these higher level behaviours are not needed they are ignored.

Genghis has been given a wide repertoire of behaviours. It can scramble well over rough terrain, cope with steep slopes and even be made to track and attack anyone or anything that moves through its field of vision.[67]

How Genghis is made to walk illustrates another of the fundamental aspects of Brooks' approach: emergence. Out of the chaos of all those conflicting signals emerges several different gaits and walking behaviours that Genghis can use to cope with a variety of terrains. If it feels the sand slipping beneath its feet, Genghis will pause and compensate before trying to move on. If it feels the ground slipping as it places a foot, it will pause, retract the foot and try again, just like an insect. This ability to adapt to rapidly changing conditions and terrains could also be taken as proof of intelligence, but there is no place within the robot's control system that you can point to and say that is where the intelligence lies. The intelligence emerges in the interaction between us and the robot or insect. We can impute intelligence to Genghis in the same way that we can appreciate that there is a crude but fierce intelligence at work within a cockroach. Neither may be as intelligent as you or I but they are not stupid either.

There is a lot of information about these robots, and others, in a paper called 'Fast, Cheap and Out of Control: A Robot Invasion of the Solar System', a title that went on to inspire a lot of badges and even a book by Kevin Kelly about those using biology to build better machines. The paper appeared in the *Journal of the British Interplanetary Society* and was not just intended as a review of past work. It was also written to illustrate the idea that it makes much more sense to use robots to explore space than it does to send humans. The life-support systems that humans need in space are bulky and expensive to get off-planet. Far better, argues Brooks, to send robots in our stead to prepare the ground for us and find out if it is worth travelling all that way.

Brooks was not the first to suggest that robots should be our ambassadors to other planets. Since the early 1980s NASA had been thinking about just such a plan, but unfortunately it had limited success in building robots that could do the job. The big problem to be overcome when controlling a robot millions of miles away is the time delay between sending a signal and the robot receiving it. There is also the possibility that the communications path between the ground station and the satellite is blocked by a planet. If a robot is heading for a precipice it is no good sending a stop signal from Earth; by the time the pictures of the looming danger have arrived on Earth the robot will probably be lying in a heap at the bottom of the chasm. Far better to send a robot clever enough to spot when it is in danger or so cheap that having one fall into a ravine is no great loss. If lots of similar robots are sent, the loss of one robot becomes easier to bear.

The idea of using autonomous robots has obviously caught on. The little Pathfinder robot that landed on Mars on 4 July 1997 was partially autonomous. Its handlers at JPL on Earth picked the rocks and features that it should look at and gave it a path to follow, but the roving robot was smart enough to work out for itself how to get there if it needed to.

NASA is planning many more missions to Mars. The majority of these missions will use robotics to explore the planet. At the time of writing NASA has not yet decided which robots it will use or what they will be doing. One of the missions involves a lander that will gather Martian soil and rocks and then send them back to Earth in another, smaller, craft. Autonomous robots will be essential to such a mission. This sample-and-return mission is likely to use robots to do the gathering of rocks, it may also use other robots to extract the fuel for the spacecraft's return journey from the Martian atmosphere. Such a plan would drastically reduce the amount of fuel that the sample–return craft has to carry with it.

It is not just planetary rovers that will be exploring all by themselves. As part of a plan to reduce the cost of long space missions NASA has been developing autonomous space probes. In October 1998 the first of these probes, Deep Space One, was launched. This small 225 kg spacecraft is due to fly by the asteroid 1192KD in July 1999 and then, if its mission is extended, tail the comets Wilson-Harrington and Borelly, in 2000 and 2001. At the same time DS-1 will be testing out software that NASA hopes will slash the cost of such missions by 60 per cent.

Usually every satellite no matter how big or small is supported by hundreds of people in ground stations who monitor the health of the satellite and tell it what to do. In contrast DS-1 will have a much smaller ground crew who will only pass on objectives to the satellite. Then the satellite will work out itself how best to meet these objectives. Its decisions will be affected by the resources, such as power, it has available and the health of its sensors and instruments. The satellite will also have the ability to refuse to carry out objectives if it thinks that the plan it has been given is unreasonable. The software used in DS-1 will be upgraded, expanded and used on other similar missions.

Know Your Territory

Hungarian-born Maja Mataric, one of the later additions to the Mobot lab, extended Brooks' early work on the subsumption architecture. She produced a robot called Toto that could navigate via landmarks that it had learnt on its earlier peregrinations. The pathfinding mechanism in this robot was loosely modelled on a rat's hippocampus – the place in the rodent's brain associated with motor control and navigation. The behaviour of the robot helped it to move around and avoid obstacles and also helped it recognize where it was. Its knowledge of these landmarks was based on what it was doing when it first came across them. Combinations of behaviours became associated with the locations and were triggered whenever the robot returned to that point. As a robot approached a location these

MARK WARD

behaviours would be activated and the location associated
with them would be called up. This also helped the robot
work out when it was close to a place it had never visited
before because it would have no memories to call on to
identify where it was.[68]

Mataric has now moved on from MIT and runs the
mobile robot lab at the University of Southern California.
Now she is working with collections of up to twenty-four
robots – a swarm that she calls the 'Nerd Herd'. She is
teaching them to cooperate on various tasks. Cooperative
behaviours emerge when robots are given clearly defined
territories and the ability to signal to each other. Robots
that can cooperate and divide up a territory equally could
be used for mine clearing or to prepare landing sites and
bases for humans when, and if, we start to settle on other
planets.

Men and Cogs

The work carried out at the MIT Mobile Robot lab amounts
to a repudiation of everything that GOFAI held sacred.
Robot makers before Brooks (with the exception of Grey
Walter, Ross Ashby and perhaps Norbert Wiener) insisted
on the opposite of everything that Brooks was now suggesting. For years robot makers had tried, and largely failed, to
make their creations collect data via sensors, use this to
construct an internal model of what was out there, test a
plan of action in this world and then carry out the plan.
They ignored life and instead simulated everything. What
Brooks considers irrelevant and wrong they saw as essential
and accurate. There is little chance of a rapprochement

between the two camps because their views of brain function are so different. Only time will tell who is on the right path. As Brooks says, 'Whether I am right or not is an empirical matter.' If we judge both camps by results so far, for my money Brooks is ahead. In less time he and his colleagues have built autonomous robots that run rings round anything that has gone before. That Brooks is now head of the AI department at MIT rather than just the Mobile Robot lab is perhaps a reflection of how far ALife has come and how quickly his ideas have entered the mainstream.

Sceptics can argue that all Brooks has done is show how easy it is to mimic the intelligence that is found in insects and that this is irrelevant to anyone interested in human intelligence, which seems to be qualitatively different to that found in insects. They could say that because Brooks and his co-workers started with lower animals and therefore had lower criteria for success it is hardly surprising that they have not been disappointed. The real work, that of tackling human intelligence, remains to be done. If those tackling this lofty goal have failed, it is only because they were aiming so much higher. But Brooks is not shirking the challenge, with his latest project he is tackling these issues head on. Brooks is making his new robot deal directly with the world, giving it a body and senses to experience the world and merging this with emergence and intelligence in a five-year plan to create an artificial, intelligent humanoid called Cog.

Like all the other robots that Brooks and his students at MIT are building Cog will deal directly with the world. In fact, it will do so rather more directly than any other Mobot creation. Cog is not just being built to mimic human intelligence, it is being built to look human too. So far it has been outfitted with a torso, an arm, neck and a head fitted with eyes and ears. The torso is mounted on hips that allow it to swivel as much as we can. The bulk of the body holds some of the gears and servos that power the arm. Eventually the body and arm will be covered with a rubber skin stippled with sensors to give Cog feedback about what its body is doing – this may impart a sense of self. The neck can also swivel to bring the eyes to bear on anything of interest. The eyes themselves move in unison and can pan and focus. The head is automatically kept level using a vestibular system similar to that found in humans, which gives us information about how our head is moving and helps us keep upright.

Cog has a brain too. A network of Motorola 68332 chips is being used to create a massively parallel processing system that will help Cog learn. Although there are not as many elements in this parallel computer as there are in the brain, it is at least more realistic than the serial computer used by many GOFAI researchers. Also Brooks does not want to re-create an adult intelligence, he is aiming for a childlike but recognizably human intellect, something like a two-year-old child. A crude talker and mover maybe, but recognizably human none the less.

Rodney Brooks has described himself as 'not a self-doubt kind of a guy'[69] but many think that even he has over-reached himself with Cog. Certainly there are a lot of people who would like to see him fail, if only for the reason that if he is right, they will be out of a job, and the research they have been pursuing for the last few years will be useless because it has been demonstrated to be just plain wrong. But Brooks has good reasons for everything that is built into Cog and for expecting that this will produce rudimentary human intelligence.

Brooks thinks that human intelligence has been hard to emulate because at least some of our mental life is governed by the way that we sense the world and get to grips with it. Leave the body out and an essential part of the picture is missing. Ignoring the body removes the tool, ourselves, we use to learn about the world. Unsurprisingly there is a relationship between what organisms do and the constraints their bodies place on them, the surprise is that no one else trying to re-create human-like intelligence has given a robot enough sensors to feel the world with. Before Cog no robot has had anything like the range of sensors or senses that humans have. Cog is severely limited compared to the average human but it is a long way ahead of all other robots. These bodily constraints cover obvious things like 'only those with mouths can munch' but also in other ways such as the way that an animal moves around. Clearly what an organism can do will affect what it considers as relevant in the world. It will tend to ignore the things it cannot change. An organism may never realize what it cannot do but it will certainly know about what it can do and divide up the world on those terms. Learning your limitations is part of discovering who and what you are.

This process of discovery does not end with knowledge about how high you can jump or how far you can see. There are those who believe, and Brooks is one, that our physical interaction with the world is the basis from which all our thought and language comes. The physical interaction with the world generates patterns in our sensory and motor systems. The soft fur of a cat and the soothing sound of its purr give us more than just knowledge about contented cats. 'The physical bases of our reason and intelligence can still be discerned in our language as we "confront" the fact that much of our language can be "viewed" as physical metaphors, "based" on our own bodily interactions with the world.'[70] Our everyday language is peppered with these phrases such as 'grasp the truth', 'pushing forty', 'see sense'. In building Cog Brooks is following the route that humans have tried and tested for millennia, using physical sensations as the basis for higher level cognitive behaviour.

This is where the need for massively parallel computers arises. The human brain works like such a computer because it works on many different things at the same time. This separation of processing power will also help coordinate Cog's separate parts – a serial processor that could handle only one input at a time would struggle with this kind of task. The smooth meshing of neck, head and eye movements demands considerable independent processing power. To ensure the movements of Cog are smooth and lifelike each part of the body needs its own processing system.

The different sensory systems of Cog will also have their own processing systems. Again this has parallels with the human brain in which memories are organized by the way in which we experience them, either by smell, touch, sight, sound, or taste. The different ways we experience things are often linked, if only because some of these memories may have been laid down at the same time. Experiments with people who have suffered injuries to their brain has established that our memory is organized by the route it takes to the brain via the eyes, ears, skin, etc. Similarly Cog will have its memories organized by the sense that acquired them. Brooks writes: 'In building a humanoid, we will begin at this sensory level. All intelligence will be grounded in computation on sensory information or on information derived from sensation.'[71]

Sensations from the outside world will not just be used to give Cog an idea of what it is experiencing. They will have a wider role to play, just as they do in humans. Some will be abstracted and generalized to give Cog an idea of space and it will get to know what up, down and near and far mean. Still others will be further abstracted or simulated internally. Again this has parallels with humans. It has been found that when we use visual imagery parts of our visual cortex are stimulated. The parts of the brain that are concerned with visual experience contribute to what we are trying to convey. Similarly Cog will be given the ability to call on experience to help it communicate.[72]

Brooks says that 1998 was to be an important year for Cog. A lot of the researchers employed on the project are due to present papers and reveal how far their work has come. By now you may have heard much more about it.

My Valet Has Exploded

A growing body of evidence suggests that if you want to be a ground-breaking robot scientist you have to start young. Grey Walter's childhood hobbies led him to abandon the classics for a career in science. Rodney Brooks built all kinds of mechanical marvels from an early age, and so did Mark W. Tilden, another of the prolific autonomous robot researchers on the planet. Now the robots that Tilden makes rival those made by Brooks and his AI lab workers at MIT for the title of most lifelike and competent artificial creature.

It was not always so. British-born Tilden says he was born building but his early efforts were not as successful as his latest attempts. He claims that he built his first robot doll out of scraps of wood when he was three. It still hangs on his mother's wall in Northampton. 'I progressed from there to a Meccano suit of armour for the family cat at the age of six. I've been building devices ever since.'[73] Cats are important to Tilden. They've taught him a lot about what it takes to be a living creature in a strange world, just by the way they have interacted with his machines over the years.

'In the field of Artificial Intelligence (AI) there's a challenge called the Turing Test.' says Tilden. 'That is, can a program convince a human that it is alive through intelligent discussion. I submit that my "Purring Test" is more important to the robotics field. If you can convince a cat that a robot is alive through motion and competence, you're well on your way to making a machine that satisfies many basic living attributes.'

It was a long time before Tilden managed to build a robot that did what he wanted it to. Motivated by memories of robot helpers in science fiction movies, in 1982 Tilden tried to create a definitive robot butler. For years he tried to make a small robot slave that would vacuum his apartment while he was engaged in the much more important business of finishing his education. He outfitted the robot with four megabytes of memory and used a Motorola 68000 microprocessor as a primitive brain. This chip was programmed to follow Isaac Asimov's fictional laws of robotics (protect humans, obey humans and then look after yourself). He added instructions that helped it avoid furniture and his cat. The robot worked after a fashion but it was 'incredibly paranoid'.[74] Often Tilden would come home and find it hiding in a corner trying to get away from the cat. The end came one night when the vacuum cleaner inadvertently sucked up a pair of Tilden's socks. These tangled around the robot's wheels, making a motor burn out. Then it blew up. Frustrated by years of wasted effort he discovered that there was one job it was perfectly suited for and one that it continues to do to this day: a hatstand.

He didn't realize it at the time but his experience with the exploding robot butler taught him a valuable lesson. It taught him that programming a robot is a very complicated thing to do if you are interested in creating a useful and halfway intelligent robot. For the program to be useful it would have to contain rules and exceptions that explained how to cope with any and every situation that the robot encountered. Even in as small a landscape as an apartment that adds up to too much work. One lifetime would not be enough to write that many rules. Nor is there yet a computer powerful enough to work out what situation the robot is in, find the rule and apply it quickly. 'No matter how big the computer, simple general problems made them fall on their mechanical butts.' In a few years Tilden had gone through exactly the same experience that it took the larger AI community thirty years to complete. Both came to the same conclusion: If you start trying to build a brain using a rule-based computer you quickly come to a complexity barrier that stops you cold. He wrote later: 'We think that just because we drive around in an autonomous goal-oriented vehicle (our bodies) that we are fully cognizant of how it works. This intellectual vanity brings us disappointment when our million-dollar AI toys turn out to have fewer real-world smarts than a goldfish.'[75]

Tilden did not forget his dream of making useful robots, but for a few years he struggled to make progress. At this time he was a robot engineer at the University of Waterloo in Canada. There he worked with simulations of industrial robots and control systems and then tried to transfer them to the real world. Once again he was frustrated in his attempts to cross from the simulated world to the real one. 'I eventually realized that it is so easy to write a simulation, but nearly impossible to actualize that simulation in real life.'

Then on 29 October 1989, seven years after he abandoned work on his butler, he attended a talk given by Rodney Brooks at the University of Toronto. The talk was all about Brooks' subsumption architecture and his philosophy of doing away with programming and instead engaging the world directly. Suddenly everything became clear. Later Tilden would say of Brooks' influence: 'He basically told me everything I needed to know. Simple, elegant solutions do exist.'[76]

The talk made sense of everything Tilden had gone through with his simulations, the failed butler and even some gruesome childhood games. As a child living on various Canadian farms he had played a game called 'spoof ball'. This involves a chicken that has just had its head cut off and a load of kids. A combination that is bound to lead to lots of trouble or lots of fun depending on your point of view. The object of the game is for the kids to keep the chicken running around a farmyard by kicking or hitting it. Tilden remembers that a headless chicken can be kept running around a farmyard for many minutes. It can easily recover from a hefty kick or a stumble and keep on running. He can't remember how the winner was decided or if they got a prize, certainly the chicken was unlikely to come out ahead. One thing Tilden did come away with was the realization that not all an animal's intelligence is held in its head. Reflexes, posture and movement contribute to keeping an animal upright and active long before the brain gets involved.

Brooks was saying much the same thing. That it does not take a brain to produce competent behaviour, you can go a long way using collections of simple/stupid components. Although much of the talk made perfect sense there was one thing about it that did worry Tilden and that was why Brooks was simulating his control system on computer rather than building it using electronics. At this time Tilden was still working at Waterloo and going through his frustrations with computer-robot interfaces. Brooks replied that the reason he used computers was because it was impossible to do it any other way. Robots built around his subsumption architecture that used only electronics would be too rigid and prone to failure. He needed the flexibility of a computer to get them going, but even so they were far simpler than conventional AI programs.

Tilden took this as a challenge and only two weeks later had done what Brooks had claimed was impossible. The first robot he produced was called 'Solarover 1.0'. All of the parts for it came from broken calculators and cassette players. It was a crude, transistor-based, light-seeking 'photovore' that his cat flipped over on a regular basis. This early design was based on Brooks' subsumption architecture.

A published design for the most basic kind of robot, called a Solaroller, that drew on Brooks' subsumption architecture was laid out in an article Tilden and one of his students, Michael Smit, wrote for *Algorithm 2.2* magazine in 1991. The design of the robot and what it is built from gives an insight into the way Tilden works. Today the method remains unchanged even if the robots have become much more sophisticated.

As its name implies the Solaroller collects power using solar cells and, once charged, uses this energy to drive a motor and turn its wheels. The first problems that the design of this robot has to overcome are the inefficiency of the solar cells and the low voltages they produce. Most commercial cells convert only a few per cent of the available light into power, producing around 0.5 volts. This is a problem because the electronics need between 2 and 5 volts to operate.[77]

Smit and Tilden overcame this problem by creating a circuit out of two transistors, a capacitor and a diode that can store power until it has enough to drive the motor. The solar panel collects power and this is used to charge up the capacitor. Once the right voltage is reached, a zener diode makes the transistors dump the power they have saved into the motor. This turns the wheels and the roller jumps forward. Then the robot rests again while enough power is collected for another spurt of motion. This circuit was the first of many to come and is now called the classic A1 Solarengine.

The Solaroller robot encapsulates what is different about Tilden's robots. First, it is built from spare parts. The components that the example robot was made from came

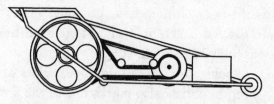

Side view of a completed Solaroller and a diagram of its electrical innards

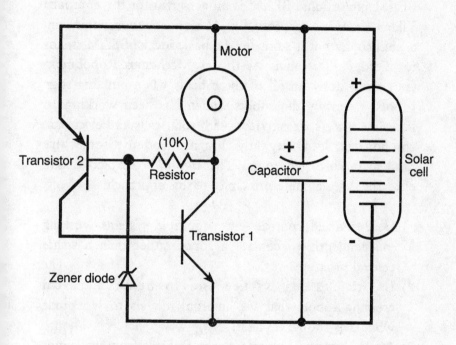

from solar-powered calculators, cameras and cassette players. The drive wheel was a section of a printer roller. Second,
the robot was autonomous. Because it used solar power it
needed nothing but the sun to get going. Third, it was
robust. Tilden says that the first Solarovers and Solarollers
are still working today, years after they were created.
Fourth, there is no computation going on at all, yet what he

has produced is an autonomous creature whose 'motions contribute toward a certain feeling that even this modest creature has a life of its own.'[78]

As well as being one of the first appearances of this new type of robot, the article also marks one of the first appearances of Tilden's distinctive approach to robot making – the BEAM philosophy. BEAM is an acronym for the approach Tilden wants it to capture. BEAM is more a work ethic than a philosophy but it sums up the basic ideals. BEAM stands for Biology, Electronics, Aesthetics, Mechanics. Robot makers should draw on all of these fields when building their robots. Certainly the robots Tilden has been working on look like insects, are studded with solar cells and electronic components, look elegantly designed and use some very basic methods to get around. The *Algorithm 2.2* piece makes clear the three guiding principles of this approach.

1) Decentralized control – use multiple systems working independently to control a robot rather than a single central program.
2) Use stimulus engines – tie sensors to effectors and avoid creating robots that use internal models to work out what they should do next.
3) Mimic biology – imitate design features from well-studied creatures, useful lessons can be drawn from biology[79] and prior robot work.

Tilden has also attached other meanings to BEAM. In the past he has claimed that it stands for Building, Evolution, Anarchy, Modularity, for example. Building because he feels that there is no substitute for first-hand experience, frustration can be a great teacher. Evolution means starting small and working up. Get something right, build it in and move on. Don't try and do everything at once. This amounts to a summary of Brooks' subsumption architecture in which new behaviours are laid over the top of what the robot already does. The *Algorithm 2.2* article makes specific mention of this approach, adding: 'Improvements should be introduced in subsequent designs, not in the present one. This prevents endless tinkering during which evolution is temporarily halted.'[80]

Anarchy does not imply a political dimension to Tilden's work, instead it is meant to imply that a certain amount of disorganization is desirable to maximize 'robogenetic' variations. Plans should not be rigidly adhered to, instead there should be room to try things out and see what happens. It is also meant to capture the autonomous, unpredictable and rebellious nature of BEAM robots.

Modularity captures Tilden's views on the relationship between what a robot does and how this is controlled. He has learnt to his cost that simulating control is a dead end. The way a robot senses what is going on and the wheels or legs it uses to react should be built at the same time to ensure they work together.

With the BEAM philosophy comes an alternative to Asimov's rules of robotics that Tilden tried to build into his failed robotic butler. Tilden's guiding principles have more in common with the words voiced by Ralph Numbers in Rudy Rucker's book *Software* and which form the extract that opens this chapter. The rules are protect yourself, feed yourself and look for better power sources; Tilden has paraphrased these as:

1) Protect thine ass
2) Feed thine ass
3) Look for better real estate.[81]

BEAM does not stop with philosophy, soundbites and wit though, there is a much more active side to it. For the last eight years Tilden has been organizing the International Robot Games, an annual competition that brings together BEAM robot makers from around the world. Like any other international sporting occasion there are many different events that the robots can take part in. Past BEAM games have included such events as the high jump, long jump, robot sumo, rope climbing and the limbo race. Typically robots are built to compete in a specific event. Usually there is also a Solaroller race in which the simplest robots compete to cover a 1-metre course as quickly as possible. The events start simple, and progress in difficulty, so that returning designers can always try their hand at more complex tasks year after year. As such, there is never a fixed ruleset, and ten-year-old amateurs have an equal chance of winning against the professionals. Occasionally such has been the case. 'A Canadian newsreporter once called this the most politically correct games on the planet,' says Tilden. 'That's something worthwhile, I think.'

While a lot of fun is had during these games, there is also a serious side to this. Because self-replicating robots are a long way off, Tilden has often said that humans are a robot's way of making new robots. The scientific reason for holding the games he says is to 'improve robogenetic stock through stratified competition and have an interesting time in the process'.[82] Mostly, though, it appears to be a forum for compulsive, megalomaniac machinists.

While it is odd to think of oneself as a robot's reproductive organs, the BEAM games are certainly responsible for introducing many new robot 'species'. Tilden himself has created hundreds of species. Trying them out against other robots trying to do the same thing, such as climb a rope or swim to the other side of a tank, is a crude and slow form of evolution, but it is still evolution for all that. Tilden's machines cannot win any prize, but they are used as 'odd man' competitors, 'pace cars', but more importantly, they are examples of functional BEAM technology built from common household electronics that he encourages people to handle and examine. Those who bring their robots along to the BEAM games go away with their heads filled with ideas about how to improve their creations for the next year.

This may be slow robot evolution, but it is evolution none the less.

The simplicity of BEAM robots does not mean that they are good for nothing outside a BEAM arena. Tilden may have failed to make a robotic butler but a few BEAM robots do the job of cleaning his apartment quite well. Tilden has several basic rover robots that drag themselves around his floor. Slung beneath the bellies of these robots are velcro strips that gather up the dust and hair on the carpet. Every so often Tilden cleans or replaces the velcro, or runs the robots through the dishwasher. This small collection of robots is called the dust-bunny cowboys because they round up all the dirt.

He also had robots in his apartment that cleaned the windows and others that killed flies. In doing this he has created a small robot ecology. As the window cleaning robot moves around it disturbs flies sunning themselves there. When they fly off some of the flies are caught and killed by the fly trap robot. The dead flies drop on to the floor and the dust-bunny cowboys clean them up. Tilden says as the robots move around he has often been struck by how insect-like they are. The ticks and whispers they make as they move around are like those you would hear on a summer day in the garden. Watching a tabletop covered in BEAM robots, some flapping their solar cell wings and others struggling for the best basking position beneath a light source, is an eerie experience. These robots look alive.

Bang Goes the Robot

Tilden has moved on from the University of Waterloo and now works for the US Government at the Los Alamos National Laboratory, building ever more advanced robots as a research scientist. He has come a long way from the dust-bunny cowboys but his approach has remained largely unchanged. One thing he has abandoned, however, and that is the architectural aspects of Brooks' subsumption architecture. Tilden says his initial tinkering has proved Brooks right and it is difficult to build the subsumption architecture out of just electronics. To do it demands a computer. But Tilden has not gone back to subsumption or putting processors on to his robots. In fact he is now building robots that do no digital computing at all, trusting more to the analog logic of pulses and values. Despite this seeming lack of sophistication he says these robots are now 'approaching competence with less transistors than a pocket radio, and not a single line of code required'.

But if Tilden has set aside the digital, computational aspects of the subsumption architecture, he has remained faithful to its other tenets. All the robots he is now building deal directly with the world, there is little or no internal model building to help them get around. All the behaviours of the robots are built up in layers.

The bottom layer is the locomotive system that controls the legs of a robot. Tilden prefers to work with walking robots because it is easier to spot when they are not working properly. With wheeled robots it is rarely immediately obvious what has gone wrong. But if the robot consistently trips itself up, it is easier to diagnose the problem. Before Tilden starts to make a robot walk it must first be able to stand on its own two, four or six feet. Legs are given joints equipped with controllers that let the robot's limbs push against each other and balance the body. The controllers use fuzzy logic to achieve this balance. Usually logic is all about working out whether a statement can be proved, whether it is true or false. Fuzzy logic allows engineers to deal with cases that are neither wholly false nor entirely true but are correct some of the time and wrong at others. This helps the legs balance because the rough terrain that the robot is traversing means the robot often has to balance in less than perfect conditions, sometimes with a leg braced against a rock, that a purely logical system would be unable to cope with.

The motors that drive the legs are also the sensors that feed into the 'nervous network' core of the robot. This core is the main difference between Tilden's approach and other conventional control architectures. Tilden has done away with the digital processor and has replaced it with a chain of transistors. These process the data coming in from the sensors in an analog fashion, rather like real neurons do. Gone are the on-off certainties of digital inputs and in its place is a much sloppier system that can deal with all stages in between. In a computer chip transistors are used like switches turning currents sharply on and off. But once freed from this regime they are capable of having a much more subtle effect. They can act like amplifiers or drains temporarily boosting or reducing a current, immediately and without processing delay.

Tilden has discovered that by connecting a few transistors together and letting them act like neurons, by provoking or inhibiting other signals, a robot can cope with a constantly changing environment. The sensor feedback from the legs of the robot provide signals into the chain of neurons. The signals they feed in change depending on what the leg has encountered. If a leg has snagged on a rock, the lack of expected input may mean that the other legs push all the harder to get the robot to drag itself over the obstacle. The constantly modified stream of signals marching around the neural chain corresponds to a dynamic picture of the world. It is the only model the robot works with. As the robot moves, this model changes as the signals flowing through the transistors change. Basically, there are no behavioural 'levels' any more. The entire robot is a single, flowing, analog computer. In one of the papers Tilden and Los Alamos colleague Brosl Hasslacher have authored detailing their approach they say: 'Here presentation is non-symbolic, consisting in the delay and shaping of a chase-series of timing pulses that drive the motors of the device.'[83]

With relatively few transistors Tilden has created robots with sophisticated gaits that can traverse rock-strewn deserts or minefields. 'We have capable walking machines that negotiate complex unknown terrain using a total of twelve transistors as the computational core.'

With careful calibration (read human-directed evolution) Tilden has given his robots a variety of different gaits, which they discover with increasing speed. A version called Walkman 1.5 was so fast on its feet that it 'has been seen to frighten researchers and various domestic animals as a consequence'. This 700 g robot can pull itself over obstacles twice as tall as it is, and lower itself from obstacles four times its height. It has the tenacity of an insect and can adapt its walking pattern to cope with many different terrains, even a pile of coat hangers.[84] All this with no programming on board.

Tilden's robots even keep going when they have had a limb or two blown off. Tilden has had a lot of his robots blown up. One of the projects he was working on for the military was the creation of a Minestomper robot that would be used to clear minefields. Tilden reasoned that the best way to find and remove a land mine is to step on it. For some months in 1996 he worked on designing robots to do just this. The final design weighed around 9 kg and was capable of finding its way across a minefield, destroying ordnance on the way. The design used strong, spindly legs that would often survive the blast of a mine going off. Even if a leg was lost, the robot kept going with only three limbs. Even when the robot was down to just one limb it usually managed to drag itself along. The Minestomper was tested in the Yuma proving grounds in Arizona, which the US Army has used as a test firing range for years and is littered with exploded and unexploded ordnance. The robot Tilden unleashed into these daunting surroundings coped with the conditions and passed the tests with flying colours. A larger final design of the Minestomper was finished and tested in

September 1996. Then the US Government re-organized the accounts of the military and Tilden lost funding for his project.

It is in areas like mine clearance that Tilden thinks his robots will come to have a use. He sees them being employed to do all the dirty jobs that humans don't want to do or can't do. 'I see them as the contents of a programmable ecology. They'll replant forests, hunt cockroaches, monitor poachers, cut your grass, clean your pool, polish your floors – all invisibly, dependably for years.'

Most recently Tilden and his colleagues have been working on robots called Pixelsats. These satellites are the size of a fifty-pence piece, weigh 14 g and are cheaper to make in their hundreds than one large satellite. Once in space a cloud of these satellites would spread out and form huge sensor arrays that bounced signals back to Earth or peered outwards at the Universe. They could surf the Earth's magnetosphere and give accurate readings of fluctuations that in the past have disabled large satellites and even caused power cuts on Earth. If some of the robots were disabled or destroyed, the others could work around the broken robots.

Tilden has a very utilitarian view of his robots and often refuses to be drawn on the implications his work has for artificial consciousness studies. He has gone on record as saying that in contrast to Rodney Brooks, he is not interested in re-creating human intelligence. 'Why would I want to build a person when I can explore unorthodox alien intelligences from the ground up?'

Despite this reluctance to speculate Tilden does have views about intelligence and where it comes from. True to Brooks' subsumption architecture he believes that intelligence comes from the world and how a robot or person deals with what it comes across. No one would call a robot that blunders around crashing into everything clever. The point here is that descriptions like clever and stupid are imposed from the outside, usually by the people that observe them. There is nothing inside a Tilden robot that can be pointed to as the seat of intelligence. It is only by the behaviour of a robot and our reaction to it that we can judge how lifelike it is. In this regard Tilden has no doubt that his robots are, if not intelligent in the classic sense, definitely competent. Again this is not because they have crossed some indefinable threshold of mental activity. He makes this judgement solely on the basis of the reactions of humans to what his robots are capable of. He says:

> I've seen the reaction of people when one of my new headed robots looks at them and tracks them around the room, or when they are chased by one of my new 6-foot-high legged prototypes. They sense it immediately; these things exhibit behaviours more like animals than periodic-motion toys. And when they see that I don't even control the robot's power let alone their actions, there is a hint of nervousness when they realize that this device looking down at them has to be treated on its own terms because it's real.
>
> What do I do now? I watch these interactions with a scientist's eye looking for clues to fine tune my research direction. Beyond the Purring test I see hints of a new measure of social interaction, the Whirring test – what does a human have to do before the robot thinks it's alive?

Chapter Four

Smaller is Smarter

Between ants and the lower forms of insects there is a greater difference in reasoning powers than there is between man and the lowest mammalian. A recent writer has augured that of all the animals ants approach nearest to man in their social condition. Perhaps if we could learn their wonderful language we should find that even in their mental condition they also rank next to humanity.

Thomas Belt. *A Naturalist in Nicaragua*.[1]

The Ants Inherit the Earth

If you were an extra-terrestrial visiting this planet in order to contact Earth's dominant lifeform and invite it to join your star-faring race, it might not be humans that you sought out first. On nearly every scale you could use to decide who is in charge, humanity comes up short. And in many cases it is the same order that beats us every time.

We are certainly not the most numerous creatures on Earth. The bacteria and insects alone outnumber us several times over. British entomologist C.B. Williams once worked out how many insects were alive on Earth at any moment, he put the figure at 10^{18}, that is one million trillion. Humans in contrast number only six billion.[2]

If you take biomass as a measure of dominance, ants rival humankind. At a conservative estimate, around 1 per cent of the total number of insects are ants, around 10,000 trillion of them. The average worker ant, by far the most abundant type, weighs between 1 and 5 mg. Multiply the two together and you get a figure that is close to or exceeds the combined weight of every human being.[3]

The social insects as a group, which includes wasps and bees, also dominate the biomass of insects. In total around 750,000 insect species have been identified by biologists. Of this total 13,500 of these species are social insects and 9,500 of them are ants. 'This means that more than half the living tissue of insects is made up of just 2 per cent of the species, the fraction that live in well-organized colonies.'[4] It is estimated that there are at least twice as many species of ants yet to be discovered than are already known about, so the social insects may be more dominant than even these figures suggest.

Ants also rival humans in their social organization. We may count our cities as our greatest achievements and proof that we dominate this planet, but ants also live in well-ordered colonies. Some leaf-cutter ant nests can contain up to eight million individuals, a figure that rivals all but the largest human conurbations. Cities like London and New York have been likened to ant heaps, but no ant colony is as badly run as the most orderly of our cities. In an ant colony every individual knows their place and what they have to do.

Ants run their nests with masterly control. The temperature within most nests only varies by a couple of degrees.

Within leaf-cutter nests there are garden chambers where food for the colony is cultivated. Other ants farm aphids for their honeydew or mealy bugs for protein. Still others form symbiotic relationships with plants such as the acacia. Ants have been carrying out such management for much longer than humans have. More evidence for this was found in January 1998 when some fossil ants were discovered that were 92 million years old.[5]

Nor can humans distinguish themselves by their savagery. Ants fight wars with far less pity and more ruthlessness than humans ever did, they are happy to take slaves and work them until they drop, and their colonies are ruled by an absolute monarch. The army ants of southern America and the driver ants of Africa have evolved into entirely nomadic species because of their rapacious appetite. When they go raiding they denude the countryside so completely that if they established fixed nests they would quickly starve. Instead their nests are made up of their living bodies and are dismantled and moved on as the food supply dwindles.

Ants also have at their disposal a formidable ability to communicate, some have said that the main reason for their success is because, like humans, 'they talk so well'.[6] Most of their messages are passed in the form of scent trails. A courtly dip of the abdomen is enough to leave a marker that every ant can understand. In a dark underground nest, scent is the only sense that matters. Above ground ants employ more senses to find their way around. Foraging ants use visual cues to remember where they found some food. They also have a limited temporal awareness and, like Pavlov's dogs, can be trained to expect food at a certain time.[7]

As others have written, individually ants are among the

least sophisticated insects imaginable.[8] 'One ant alone is a disappointment; it is really no ant at all.'[9] Certainly individually one ant is no match for one man. One army ant has around 100,000 neurons, the human brain has 100,000,000,000. Collectively though it is a different story. An ant colony that is made up of 500,000 workers has a mass of neurons that is half the size of a human brain. Communication between the elements of the ant colony may be slower than that in the brain yet few would deny that a nest of army ants has a formidable array of behaviours at its disposal. It might be the case that human brains work along the same lines as an ants' nest, with many individual elements working together out of which 'intelligence' emerges. Marvin Minsky says as much in his classic work *Society of Mind*.[10] Perhaps the brain is a kind of superorganism and intelligence is nothing more than the froth on top of a lot of individually unsophisticated elements.[11] It is a point of view that is shared by many in the ALife field but one that few of them would admit to. Thomas Belt, the author of the book that I took the opening quote from, was not so coy with his opinions. After watching an army ant raid, observing ants freeing a comrade that Belt had trapped in a ball of mud, and seeing his garden destroyed by leaf-cutter ants, the Victorian naturalist wrote:

When we see these intelligent insects dwelling together in orderly communities of many thousands of individuals, their social instincts developed to a high degree of perfection, making their marches with the regularity of disciplined troops, showing ingenuity in the crossing of difficult places, assisting each in danger, defending their nests at the risk of their own lives, communicating information rapidly to a great distance, making a regular division of work, the whole community taking charge of rearing of the young, and all imbued with the stronger sense of industry, each individual labouring not for itself alone but for all its fellows – we may imagine that Sir Thomas More's description of Utopia might have been applied with greater justice to such a community than to any human society.[12]

Possibly because of their resourcefulness and energy ants have become the unofficial mascots of the ALife field. This may also be due to the fact that a lot of what they do can be captured on computer. It is easier to mimic the actions of a collection of unsophisticated creatures like ants than it is to replicate the waggle dances of bees or the behavioural repertoire of wasps. This approach of using many small self-contained programs to carry out a task has become known as the agent approach and is proving a fruitful avenue of research for scientists working in such diverse fields as evolution modelling and the control of telecommunication networks.

Many of the leading lights of the ALife field have at some time modelled the behaviour of ants, but with one exception few ALife researchers have ever actually got their hands on real ants. The one man who has, Tom Ray, has been inspired by his work with them to create one of ALife's most lasting contributions.

In his undergraduate days Ray did field work for one of the most renowned myrmecologists of the 20th century, Edward O. Wilson, who has spent much of his life studying leaf-cutter ants. Ray found that he had an aptitude for digging out the heart of a living colony that contained the queen, her nurse ants, and an ant-garden that had enough food in it to keep the ants going while they were being transplanted.[13]

While working in the Costa Rican rainforest with Wilson, Ray made some other discoveries. While watching army ants on raids, Ray noticed that, like any army on the move, the raiders were accompanied by a wide variety of camp followers. As an ant raid progresses, insects rush ahead of the van, trying to escape being eaten. Winged insects take to the air to try and fly to safety, a tactic that is likely to fail because of the antbirds that follow the raiding ants and pick off the airborne insects. Following the antbirds are ant-butterflies that feed off the nitrogen in the antbird droppings.[14]

Some of the first papers Ray published were on this relationship between ants and the hangers-on. Others have revealed that the relationship is deeper than even Ray thought. Nigel Franks has found that many more insects depend on the army ant colony for their survival. When an army ant colony moves a diverse collection of camp followers moves with it. The ants tolerate as guests 210 species of rove beetle, 10 species of millipedes, 200 species of phorid flies and a variety of wasps and mites.[15]

Dead End Discovery

The discoveries Ray made in the Costa Rican rainforest served to confirm his interest in ecology. To this day he maintains a connection with Costa Rica. He owns a house at Finca El Bejuco in Costa Rica and is establishing a 30-hectare nature reserve around the house. The building is

half-home half-lab and Ray still spends time there researching the diversity of the rainforest.

In particular he was curious about the ways that every
living thing seemed to depend on everything else. He was
fascinated by the complex relationships that had evolved
between the army ants and the many species of birds and
butterflies and other insects that followed the marauding
colony.

This fascination motivated him to complete a Masters
and Doctorate in ecology at Harvard University. His thesis
concerned the tropical vine *Monstera gigantea*, but while he
was studying ecology he made an equally profound but
much more depressing discovery – ecological theory in the
1970s was no help when it came to revealing how natural
co-dependent relationships, like those he witnessed in the
rain forest, could arise. Like Wolfram, Ray realized that a
lot of biology, and even more of ecology, was just history.
All its practitioners were interested in was documenting
what individual animals did and when they did it, not why
they worked together or how such co-dependent relationships might arise. No one seemed to be interested in teasing
out how evolution works and how complex ecosystems
emerge. Ecology was all description when there was so
much explaining to do.

His frustration with the inability of ecology to explain
what he had seen in the Costa Rican rainforest was matched
by his realization that biology derives its theories from the
weakest possible case. It only considers one instance of life
– the teeming hordes that have evolved on Earth. Every
theory of biology is derived by observing or dissecting
carbon-based organisms. Completing a case study with only
one subject is not enough evidence if you want to try and

expose the fundamental workings of life or consider it on evolutionary time scales. What was needed was a better way to study evolution. Later Ray would write: 'The ideal experimental evolutionary biology would involve the creation of multiple planetary systems, some essentially identical, others varying by a parameter of interest, and observing them for billions of years.'[16] Years later Ray has managed to create his experimental worlds, but they do not hang in the heavens. They are much closer to home and anyone can visit them. They exist inside computers but they are not just another program. Some have argued that they are a new form of life.

Part of the reason that Ray chose to use computers is because of a casual comment he heard when he was studying ecology at Harvard during the 1970s. At that time he was an avid player of the Chinese game of Go and used to attend a club for enthusiasts in Cambridge, Massachusetts. One evening his opponent was one of the pony-tailed hackers from the nearby MIT AI lab with whom he analysed the game in biological terms. Then the hacker calmly mentioned that computer programs could also be made to reproduce. Ray asked how, only to be told that it was a 'trivial'[17] matter to make a computer program replicate. Ray cannot now remember what the hacker said when he probed further, intrigued by the idea of computer programs breeding. Since then he has tried to find out just who it was who enlightened him but his search has been fruitless. 'I tried, but could never track him down,' he says ruefully.

The hacker's comment stayed with Ray and helped him when he was looking for novel ways to study evolution in the raw. This was at a time when the newspapers were full of dire warnings about computer viruses and how plagues of them were threatening to overwhelm computers and bring life as we knew it to a ruinous end. The potential of rogue computer programs to cause harm was being debated partly because over the days of 2 and 3 November 1988, Cornell computer science student Robert Morris nearly crashed the Internet when he released a 'worm' program on to it.

A worm program replicates like a computer virus but causes no harm to files, it just finds the links a computer has to other machines and creates more and more copies of itself. Morris's worm made so many copies that many of the 6200 computers the 99-line program infected were overwhelmed. The program managed to spread itself so far by exploiting a loophole in a common e-mail program. Morris designed the program to run as an invisible background task that no one would notice. Unfortunately at the time he let it loose there were still bugs in it and it replicated many, many more times than Morris expected.

Ray was not the first to see that all the talk of reproducing, mutating programs revealed an affinity between computer viruses and living things. Other researchers such as Gene Spafford are still researching this topic,[18] but Ray was one of the first to see that computer programs could be used creatively. A properly written program would be a useful scientific tool and might help him study naked evolution.

By this time he had become an assistant professor at the University of Delaware with a grant to study tropical ecology. The possibilities of studying evolution directly using computer programs proved irresistible and he neglected his duties to pursue the idea that possessed him. He kept at it despite the dismissive reactions of his students and colleagues when he told them about his ideas. He remembers bringing up the idea at a graduate seminar on ecology and nearly being laughed out of the room. But he persevered and spent a few feverish months studying computer manuals to learn how to create the program he needed. This was a skill he had to possess if he was to be able to create and seed the artificial worlds.

Not everyone was as sceptical and an inquiry placed on the Internet for like-minded researchers put him in touch with Chris Langton and others in the emerging field of artificial life. He attended the first ALife conference that was held at the Los Alamos National Laboratory in September 1987. Meeting other scientists doing research similar to his convinced him that there was worth in what he was doing and kept him at it.

Tierra Boom Day

The problem he had set himself was formidable. What he wanted to create was a computer program that would let him model open-ended evolution directly. As it ran he would gain an insight into how evolution works. Ray knew that many people were using genetic algorithms (GAs) to harness evolution, but he did not want to use them because he felt they were too restrictive. Crudely put, GAs select the best performing programs out of a mutated population. The programs in the population are made up from pre-defined chunks that carry out specific tasks. These chunks correspond to genes that together make up the programs' genome. Useful as this is proving to be in many situations (see Chapter Five) Ray considered the pre-defined nature of the GA genome to be a fatal flaw. To him GAs were a closed system that did not encourage novelty. All they did was produce one high-performing program and Ray was interested in the rest of the cast. What he wanted was an open-ended system in which evolution could be left to run rampant. Nothing should be designed in and only evolution should be directing the action.

Ray was wary of GAs also because of the way that they handled reproduction. The programs produced in successive iterations have no independent ability to reproduce. Who lives and who dies is decided externally. Every fresh batch of programs is inspected to see how well it does the job the GA has been written to tackle. The best performing programs (those with the highest fitness functions) survive and the underperformers are deleted. Again Ray saw this as a limitation he wanted to avoid. 'Self-replication is critical to synthetic life because without it, the mechanisms of selection must also be pre-determined by the simulator. Such artificial selection can never be as creative as natural selection.'[19] He adds: 'Simulations constrained to evolve with pre-defined genes, alleles, and fitness functions are dead ended, not alive.'[20]

Part of the reason that GAs impose these limitations is because of the inherent brittleness of computer programs. One wrong command or character can mean a program comes to a crashing halt. To cut down the numbers of unworkable programs GAs limit the possible variations. Removing these restrictions and letting computer programs mutate freely would seem to be more a recipe for producing lots of useless programs rather than the best way to study evolution in the raw. As Ray himself says: '. . . the ratio of viable programs to possible programs is virtually zero.'[21]

Ray got around this problem by creating the equivalent

of another computer inside his desktop PC in which the dramas of the new world could be played out. This 'virtual computer' has its own programming language so software written to run on it will not run anywhere else, eliminating the risk of anything harmful escaping. Ray has dubbed this virtual world Tierra (Spanish for Earth). Walling off his organisms in the equivalent of a game reserve has proved useful for several reasons.

Firstly, it allows Ray to create a more forgiving environment for the replicating programs. Here nonsensical instructions produced by mutating programs will not cause a program to crash. If they fulfil no function or are unintelligible they will be ignored. These useless lines of code do still have a role in determining the fitness of the organisms in question but they will not stop the program working.

Secondly, because the little programs can exist only in their specially designed game reserve there is no danger that they will escape to cause havoc on the Internet as their cousins the computer viruses are threatening to do. Ray brought in this safety measure on the advice of Chris Langton and another ALife alumni James Doyne Farmer. Ray is well aware of the dangers that self-replicating programs pose, '. . . the potential threat of natural evolution of machine codes leading to virus or worm types of programs that could be difficult to eradicate due to their changing "genotypes".'[22]

Thirdly, because Tierra is a virtual computer it is not tied to the generation of hardware that it was written on. This makes it easier to move it to newer, faster computers. It has also helped Ray and colleagues extend the range of environments available to the evolving organisms and give them much more room to play in. In these virtual worlds the organisms do not compete for sunlight and food but for resources just as essential to their continued survival: CPU time and memory space. In Tierra CPU time is the equivalent of energy and computer memory is the equivalent of territory. Successful organisms manage to occupy a progressively larger proportion of the 60,000 bytes of RAM set aside in the host computer for the Tierra world. Evolutionary success is measured by the proportion of memory space a species manages to fill. Successful organisms would also have more and more of the instruction inside them processed by the computer. The programs or organisms themselves are made up of instructions and Ray designed Tierra so that the computer looks at each line of the program in turn and tries to carry out the instructions it finds there.

If left unchecked the organisms in Tierra would choke the world and stop it working. To stop this happening Ray introduced a 'reaper' program. Periodically this creates more living space by weeding out poorly performing or very old programs. Organisms that have little or no beneficial mutations and do nothing are weeded out quickly. But even if a program is successful and survives many generations, death will come to it in the end. The reaper is first activated when the Tierran landscape is 80 per cent full.

Although Ray was at pains to make Tierra as portable as possible and able to run on other computers, the raw material he used to create the first version was the set of instructions for the Intel 80X86 family of processors. Ray cut this instruction set down so it was of the same order of magnitude as that used by living things in which information is encoded into DNA.[23] To do this he removed all the instructions that use numbers to work out which line of the program they have to look at next. This reduced the number of instructions down to thirty-two but introduced another problem, it meant that all the programs produced using these instructions would have to do all that they did line by line. There was no way to jump to other parts of the program that might be useful in certain circumstances. Removing the numbers meant that there was no way of moving swiftly to loops or branches the programs/organisms created within themselves without first executing all the lines of code between the instruction and the branch. Because Ray was trying to produce programs that could innovate and adapt this was a serious limitation.

He got around the problem by creating two new instructions that by themselves did nothing. One instruction stood for binary one (1) and the other binary zero (0). They became active when prefaced by another new command – the JMP (jump) instruction. Then they were treated as binary numbers. If a computer processor encountered a 'JMP' followed by a series of the 0 and 1 instructions it would look throughout the rest of the program for the corresponding digital number. Then it would carry out the instruction following this binary number.[24] To ensure looping and branching did not all go one way Ray also created a jump backward instruction that makes the computer processor look towards the start of the program for a complementary number.

Organisms in Tierra are computer programs made up of a collection of the thirty-two possible commands in the Tierra instruction set. Each instruction is a five-bit number, so a whole program is a long list of these numbers. The instructions act as DNA for the Tierran world and those that an organism contains are its genome.

These organisms can be made to mutate in three ways. First, single bits within instructions can be flipped, producing another of the instructions in the set or possibly a command that does nothing. Ray likens this to mutations caused by high-energy cosmic rays. Secondly, mutations also occur when organisms are copied between locations in the chunk of RAM that makes up the Tierran landscape. When the program is set running the number of mutations introduced during copying is specified. In this way programs can change and new organisms can emerge. The mutations may produce a novel organism that reproduces faster than its parent or perhaps a mutant that cannot evolve and which soon dies out. Thirdly, organisms can mutate when the instructions they are made of are carried out. Occasionally an instruction will be misinterpreted and the computer will try to do something with it that it was not supposed to. Finally, all Tierran organisms can indulge in parasitism. They can, if they possess the right instructions, choose to read and use the code of another organism. In this way even organisms that are not of themselves very successful can prosper.

Useful as this proved, some have suggested that it causes more problems than it solves and dilutes the explanatory power of Tierra and its applicability to real life.

Ray says he is not interested in re-creating the evolution of life itself but what happened once it got going. He is less concerned with mimicking pre-biotic conditions on the early Earth than with how organisms diversify once they have evolved. The event he wants to model eventually is the Cambrian explosion 570 million years ago when multi-cellular animals first appeared and swiftly colonized all the available ecological niches. In Tierra he hopes to be able to see the same forces that shaped that event at work. Ray does admit that the first runs of Tierra, before it was expanded to run across the Internet, are more akin to the early RNA-filled seas of the Archaean Earth than it is to conditions immediately prior to the Cambrian explosion.

Because he was not interested in the origins of life Ray decided that the first organism that would be inserted in Tierra would be handmade, Ray did not want to have to wait for it to emerge. This 'ancestor' organism was eighty instructions long. Ray planned to fill the world with this organism and then let it reproduce and see what evolution made of it.

Early in the morning of 3 January 1990 Ray was ready for a trial run. He assumed that although he had his ancestor organism, it would not be sophisticated enough to use as a test subject, all it could do was reproduce. In fact he intended to use this ancestor to debug the Tierra program[25] and thought it would take years to get evolution out of the system. As it turned out he never had to create another creature until he came to start work on the Internet version.[26]

Ray let the ancestor loose in Tierra and sat back. The early versions of the program ran at twelve million instructions per hour. Later versions that ran on university computers would run at speeds in excess of one million instructions per minute. The ancestor program takes over 800 instructions to reproduce so Ray thought he was in for a long wait. He kept an eye on an indicator that showed what length organisms were currently living in Tierra.

Initially, everything was exactly the same size as the ancestor. Then, as the mutations began to affect the organisms, the mix changed. The first new organisms to appear were only seventy-nine instructions in length and because they managed to reproduce faster than the ancestor they quickly took over as the dominant organism. As the numbers of the 79-creature increased and those of the 80-instruction organisms dwindled, other even smaller organisms started to appear. When he realized what was happening Ray was amazed. This was exactly what he wanted to see but he was seeing it much quicker than he thought he would. Evolution was throwing up new creatures that were better fitted to occupy the space available to them. Creative solutions were emerging, forced into being by the changes in the Tierran world. As the environment filled up with organisms it became a very different place to live. No longer was it an open landscape where all niches were available to colonize, it became a world where those that adapted best survived and many of the others just hung on. Abilities that were useful initially became replaced by aptitudes that were needed when competition was fierce.

Ray noticed that the numbers of one creature that was only forty-five instructions long were increasing. Prior to setting Tierra running, Ray had thought that it would take around sixty instructions for an organism to reproduce, forty-five seemed too low. When he later came to examine the organism in depth he found that a random, lucky mutation had created a parasite. To survive, the 45-instruction organism had found a way to latch on to a larger host that had the self-reproducing instructions it needed and borrowed them when it came time to reproduce. This was not the end of the diversity that Ray witnessed in Tierra. The artificial world has created organisms that made themselves immune to parasites by hiding the reproductive instructions. Soon after the immune strain emerged, other creatures evolved that could circumvent this immunity. This kind of push-me-pull-you arms race is well known in natural ecologies where one organism temporarily gains an advantage over another. Many of the protection mechanisms such as shells, camouflage and poisons are thought to have been provoked in this way.

After running billions of Tierra cycles Ray has found that all manner of bizarre creations have emerged from the Tierran world. He has catalogued hyper-parasites that are even smaller than the parasites and exploit their free-ride tactics. Ray has collected data on all the different types of creatures that have emerged from Tierra and has catalogued over 29,000 genomes that come in 300 different sizes. Many of them make much better use of the resources of the Tierra world than the ancestor that Ray wrote.

The diversity of Tierran fauna is not the only lifelike characteristic that this artificial world possesses. When Ray came to chart the rise and fall of populations he found that the pattern it created looked very similar to that found among real populations. Long periods of stability in which numbers and types of creatures hover around one value are broken up by short, furious bursts of evolution. During these times populations and species diverge wildly until some organisms emerge as dominant and complex ecologies become established. Ray saw exactly the same thing in Tierra. As the instruction cycles clicked around, one size of creature would tend to dominate, then abruptly mutants that were wildly more successful would appear, populations would fluctuate, some organisms would teeter on the edge of extinction, many others would die out all together. Then the numbers would settle down again. This phenomenon is called punctuated equilibrium, or more prosaically 'punk eek', and it is believed by some palaeontologists to be at work in the evolution of life on Earth.

There is no doubt that Tierra should be counted among

the first rank of ALife achievements. Some have said that it 'is likely to rank as one of the most important developments in twentieth-century theoretical evolutionary biology.'[27] They are moved to make this claim because, for the first time, it offers biologists, or anyone who is interested and has a computer, a way to study evolution directly. Evolution on demand is there for all to see. Prior to Tierra all the suppositions were just that. Now we have a world ready to tinker with that can perhaps help explain how we got here.

Others, however, have pointed to its shortcomings and have been wary of drawing too many conclusions from it. Most of all they complain that Tierra is not lifelike because all the organisms it produces are made up of fewer instructions than the original ancestor. One trick evolution has learned, but Tierra has not, is how organisms slowly bootstrap themselves into larger creatures, into tissues rather than selfish cells. Ray agrees and has been working on a way to overcome this shortcoming.

Net Gain

As Ray himself admits, trying to create the conditions for multi-cellular life to emerge is difficult. After all, life on Earth was content to stay unicellular for over three billion years before blooming into giddy diversity. So as others have pointed out, 'multicellular life of modern design occupies little more than 10 per cent of Earthly time'.[28] That it took the huge laboratory that is Earth billions of years and many times that number of false starts to produce multicellular organisms suggests it is a tough problem. But Ray is undaunted.

As Ray saw it multicellular life has three basic features that would have to be captured on any accurate computer model. Firstly, multicellular organisms started out as single cells that reached their more complex form using binary fission. Ray saw no problem turning the small chemical analogues used in the early version of Tierra into larger discrete cells that he hoped to encourage into multi-celled cooperation. Secondly, multicellular organisms have the same genetic material as the original cell. Ray had seen Tierran instructions survive between generations just as DNA does. Thirdly, although the different cells on an organism have the same genetic material they express different parts of it and carry out separate tasks. With a little clever programming Ray thought that this too was a property that could be captured.[29]

Ray also had to find a way to simulate another process that is found in the living world but rarely in computers. The legacy of the von Neumann architecture means that most computers work in serial fashion. Because they usually only have one microprocessor they can only process one instruction at a time. Living things are the opposite. In effect every cell is a little processor whirling away to solve its own problems. Admittedly many cells work on the same task in organs and work more efficiently as a result, but most are selfish in the tasks they take on. Ray had to simulate this because there were not enough parallel computers around for network Tierra to run on.

A previous ALife researcher did try to build hardware that was specially designed to support simulations of living systems. In the early 1980s Danny Hillis founded a company called Thinking Machines that made computers, called Connection Machines, with 64,000 processors, each one of which could be used to support an individual cell or entity. MIT graduate Hillis wanted to use these to do pure ALife research but commercial realities intruded. The company was formed to find other ways to exploit the supercomputers the company was making.

The most lasting contribution to ALife that used the Connection Machines was carried out by Karl Sims. He evolved a bizarre collection of polygonal creatures that flapped, bumped and hopped their way around an artificial landscape.

Unfortunately, although Thinking Machines researchers were making significant contributions to ALife, the company's commercial success was limited. In 1995 Thinking Machines filed for Chapter 11 bankruptcy protection.[30] The dream of using the supercomputers the company made for ALife research crashed at the same time. Now the workers who stayed develop software to run on other supercomputers rather than the ones the company used to build.

Sex, Sexes and Success

Just as in the original Tierra Ray created the ancestor organism himself. This was more of a challenge because no one knows what made single cells join together to form multicellular organisms. By the time he had finished, the Tierran instruction set had doubled in size and he was forced to compromise his original vision to do the job.

One of the first new instructions he wrote was the 'split' command. This allows a cell to divide into two, giving each its own slice of processor time and is the analogue of binary fission. This allows cells to proliferate and grow from a single cell into tissues. Another instruction called 'csync' coordinated the processing times that the same cells got, to make them work in harmony and act as one.

Ray also introduced sex into Tierra to try and encourage an increase in complexity. To do this he drew on work done in 1992 by Daniel Pirone, who had tried to turn Tierra into a engineering tool to breed computer programs. This gave the Tierran organisms the ability to indulge in cross-over or haploid sex, the kind that any species with different or reproducing types (males and females) indulges in.

Using genetic operators introduced unforeseen complications. He also had to add a mechanism that preserved the instructions relating to sex or very quickly the population would lapse back into the binary fission-type reproduction that Ray used in the earlier version of Tierra. Ray admits that this represents a philosophical shift that he has resisted before. In the first version of Tierra he wanted the cells to work everything out for themselves. He defends the change by saying: 'The goal I am after is for evolution by natural selection to cause an increase in the number of cell types . . . in order to be successful, it must be evolution, not me that caused the increase.'

Ray also needed a way to create those separate worlds with different conditions that would act as selective pressures and provoke evolution. This, he thought, might encourage the creatures to develop in different ways. By the time he was considering doing this Ray was working at the ATR labs in Japan and Tierra was famous, becoming well known within the ALife community and the wider world. By using the computational resources of this international community Ray saw a way to mimic different environments. Many people wanted to help Ray push Tierra further and most of them worked at academic institutions with large computer systems on site connected internationally by the Internet. What Ray wanted to exploit was the unused computer time at these places. Over the course of a day the computational resources available at one site would dwindle as those on another became free. Ray thought that this might be just the kind of active environment needed. As one campus was shutting down for the day, at another students and teachers would be waking up and checking their e-mail and at a third in the middle of the day when everyone was working there might be little processing power to spare. Reasoning that this daily flux of available computer power might prove a powerful force for selection, Ray enlisted the help of these interested researchers to create a network version of Tierra. Organisms unleashed into this new version would have a variety of worlds to choose from and a cycle of feast and famine that seemed more akin to the real world. Perhaps nomadic species of organisms would evolve that flowed across the globe as night fell and more computer power became available, allowing them to feast while people slept.

Before Tierran organisms could be unleashed on the network Ray had to give them sensory equipment to spot which habitat was best. He created a series of instructions that reveal to a cell five key features of the locations spread around the Internet:

- how densely populated the location is;
- how large a chunk of memory has been set aside for Tierran cells to inhabit, how much living space there is;
- how many instructions have been run on that habitat,

this reveals the age of the location and gives clues to how hospitable it is;
- how easy it is to get to that network node;
- the time it takes to travel to the new node.

In June 1997, Ray and colleagues were ready for a trial run. Rather than attempt to run the test over the Internet, the researchers decided to use the small local area network (LAN) that connected many of the computers in the ATR labs. The computers in many offices are wired up using similar networks. By keeping the test local the team was able to spot problems quicker. Although the test worked, Ray found that nearly 90 per cent of the creatures that migrated to greener pastures did not reach their final destination. Evolution did occur but the fatalities associated with mobile organisms was a 'strong selection pressure against migration, and it could be seen that creatures rapidly evolved to reproduce only on the local node, and not migrate over the net'.

Acting on a suggestion made by Manor Askenazi at one of the International Tierra workshops, Ray has now implemented a measure to encourage mobility and punish static organisms. Every so often an apocalypse will strike one of the nodes in the Tierra network wiping out all the cells it finds there. In this way Ray hopes to produce an increase in complexity and diversity. The most striking difference between network Tierra and the stand alone version was that the LAN tests produced far less variation in the size of organisms. Ray put this down to the low evolvability of the larger instruction set. The original ancestor was only eighty instructions long but the network version was far bigger and in the end occupied 320 bytes of memory space.

Testing continued during the later months of 1997, more bugs were discovered and removed, and in November more network tests using sixty machines at ATR were carried out. During this test the sensory system survived through prolonged periods of evolution.

Ray says other encouraging signs are emerging. Within organisms genes are being copied and both copies of that gene are being expressed. Ray expects that the genes will now begin to change and take on new roles – a process that was thought to have taken place in living organisms when multicellular life arose. The latest runs are showing that different cell types are starting to appear and work in tandem. Ray feels he is tantalizingly close to a critical point like the Cambrian explosion.

Imitations and Limitations

Tom Ray is not the only ALife researcher to try and capture open-ended evolution in a computer program. Around the same time that Ray was reporting the first results from the early version of Tierra others, such as John Holland and Kristian Lindgren, were coming up with similar ways to study the same problem. Unfortunately others have found that all of them share another property that reduces their predictive ability. The dynamics of all the model evolutionary systems fall far short of those exhibited by life on Earth.

Holland, the man who invented genetic algorithms (see Chapter Five), developed an artificial world called Echo, which, he writes, 'provides for the study of populations of evolving, reproducing agents distributed over a geography with different inputs of renewable resources at various sites'.[31] Agents or organisms within Echo possess a set of chromosomes that are built up using the resources dotted around the computer landscape. Each agent has the ability to attack, defend, trade and reproduce. Despite the simplicity of the model Holland claims it has produced agents that display many lifelike properties. In Echo he has seen mimicry and biological arms races take place as well as immune-system responses and the emergence of trading relationships.[32]

Lindgren tackled open-ended evolution in a way that was completely different from that chosen by Ray and Holland. Lindgren concerned himself with the prisoner's dilemma. The origins of this problem are disputed, some claim it was thought up by Melvin Drescher and Merrill Flood at the Rand Corporation in 1950, others claim that mathematician Albert Tucker invented it. Still others see it as an outgrowth of the work done by von Neumann on game theory in the 1940s. Whatever its origins, since its creation it has been used in many social psychology experiments. It also serves as an easy shorthand for the kind of choices biological organisms face because it is a simple demonstration of the pitfalls and benefits of cooperative behaviour.

The game concerns two men who committed a crime together and have been caught. They are being held in a police station in separate interrogation rooms and cannot communicate with each other. Alone each prisoner must decide whether to own up to committing the crime or cooperate with the other prisoner by keeping quiet. The severity of the punishment they receive depends on what, if anything, they decide to reveal. They will both be fined if they both admit their guilt. If they both maintain a stony silence, they will both go free. If one confesses he will go free and be rewarded and the other prisoner will be punished severely. Cooperation and trust get their rewards, but equally betrayal might be just as productive. The dilemma comes from knowing whether or not your partner can be trusted.

If the game is played just once, then betrayal is probably the best tactic.[33] But if it is played many times it becomes a very different game because the other prisoner's tactics need to be taken into account. In such a situation the game can be taken as a metaphor for the relationships that emerge in business or biology.

A 1980 study by Robert Axelrod pitted fourteen separate strategies against each other to see which emerged as the best. This showed that a tactic called Tit-For-Tat developed by Anatol Rapoport worked best. Whatever one prisoner did the other would do the same in the next turn, cooperation was matched by cooperation and betrayal by betrayal. It encouraged cooperation because it made the prisoners think of what may happen in future games, rather than just in the here and now. It was better to encourage cooperation because in the long run both benefit most from this strategy.

When Lindgren played the game he used 1000 dumb agents and let it run for tens of thousands of iterations on computer. He also added 'noise' to the mix, so that every now and again one of the strategies adopted by an agent would change randomly. He found that with many players and thousands of games the dilemma starts to resemble a complex ecology. Co-evolution and symbiosis emerge but so does parasitism. When the noise is turned off the stable strategy picked by one agent will be exploited by others to their advantage. Further, the numbers and types of strategy picked fluctuate just like the populations of living animals. Punk eek made another appearance and life appeared to be mimicked successfully.

Others have created models that are closer to Tierra and because of this have produced similar results. Bugs uses agents on a grid that can forage, feed, reproduce and die. Evita is similar to Tierra because its organisms possess self-replicating strings of computer code.[34]

Finally, ALife researcher Christoph Adami at the California Institute of Technology produced a version of Tierra called Avida that tries to make the simulation of populations of cells much more realistic. Avida abandons the shared memory space of Tierra and plumps for a cellular automaton-like grid on which the cells live. Coupled with this is the stipulation that cells can only affect other cells in their immediate neighbourhood. One of the criticisms levelled at Tierra is that any cell can affect any other cell no matter where it is in memory. All the locations are effectively next door to each other. In Avida the landscape is more realistic and information can only move around locally. It also helps the population survive infection by parasites. When every location is only one step away viruses find it very easy to get around.

In other ways Avida is very similar to Tierra. It uses a small set of instructions and employs a reaper to manage population explosions. Perhaps because of this it produces very similar results to those achieved by Tom Ray. The population of cells grows and diversifies while parasites manage to prosper on the margins.[35]

Taken together one would think that all these models represent compelling evidence that the fundamental motivations of biology are being captured. But when they are compared directly with the real world they are found to be pale imitations at best and completely unrepresentative at worst.

Adapt or Die

The comparisons have been made by Norman Packard and Mark Bedau. Using evidence drawn from the fossil record they have generated statistics and trends against which the claims of ALife worlds such as Tierra, Bugs and Evita can be measured.[36]

Packard and Bedau have attempted to measure the adaptive success of the organisms found in the fossil record and alive today. Simply measuring diversity was not enough because the numbers of different species dwindles as evolution rolls on. The tree of life is constantly thinned out by evolution. The Cambrian explosion may have spawned huge numbers of novel creatures but many of them died out quickly for no other reason than they were unlucky or were hampered by crippling adaptations. As Stephen Jay Gould puts it: 'The history of multicellular life has been dominated by decimation of large initial stock, quickly generated in the Cambrian explosion. The story of the last 500 million years has featured restriction followed by proliferation within a few stereotyped designs, not general expansion of range and increase in complexity.'[37]

Bedau and Packard are concerned with measuring evolutionary activity, specifically the novel innovative adaptations that make a difference and continue to be used. In the fossil record they took the appearance of a new taxonomic family as evidence of an innovative adaptation. The persistence of this family is taken as a crude measure of the evolutionary success of the adaptation. As Bedau says: 'If something is lasting for a long time then it's continuing to be successful.'

The graphs produced when these statistics are plotted are

revealing. They show that over time the organisms that remain are increasingly successful. A long-term trend of adaptive success is set early on and continues to rise. Bedau and Packard subjected Evita and Bugs to the same analysis and found these trends missing.[38] In unpublished work Bedau has found the same with Tierra, Echo and Lindgren's model. In all of these models adaptive success peaks early on and stays constant for the life of the simulation. Bedau even ran the models through many times more iterations than the original researchers did. Even then the story was the same, adaptive success started low, climbed to a plateau and then stayed there. Bedau has no doubt that the network version of Tierra will exhibit the same trend. He says because it is a larger-scale simulation than any of the others it may take longer to settle down, but he believes that the pattern of adaptations it throws up will resemble other models more than it resembles real life.

Bedau thinks that what is preventing these models from resembling the real world is the fact that they program in the adaptation they want from the start. They are not as open-ended as they would like to believe. 'The problem is one of having the programmer play God and settling *a priori* what kind of evolution can happen,' he says. 'In the real world you start with molecules and end up with you and me.'

But Bedau is not downcast and has not carried out his work in order to wreck the dreams of ALife researchers. 'There is a problem with this but it is not an insoluble problem,' he says. He does not think that ALife as a research discipline is bogus, or that producing artificial lifeforms is impossible or miscast. This is not to say that ALife research is not in for a shake-up. Bedau thinks that soon people will begin to realize the shortcomings of the models that they have created to date and go back and re-write them to make them genuinely open-ended.

Even the most recently developed models of evolution seem to be suffering the same problems. One such, Cosmos, is a Tierra-like system that was created by Tim Taylor, a PhD student in the University of Edinburgh AI department. He took the original Tierra aims and wrote a system that he thought would be free of the limitations, as he saw them, of Tierra. Despite this Taylor is honest enough to admit that he has not got much closer to mimicking real evolution. In fact his experience with Cosmos has left him wondering, like Bedau, if there is any future in building such models.

Cosmos is an acronym for COmpetitive Self-replicating Multicellular Organisms in Software. The name captures what Taylor was attempting. The success of Tierra may have inspired Taylor to create Cosmos but he has not simply settled for imitation. Rather he tried to be more faithful to Tom Ray's original vision, in order to rebut some of the criticisms levelled at it and show the way forward.

Just as in Network Tierra, in Cosmos Taylor is trying to model multicellular organisms. To do this he has employed object-oriented programming techniques. This is an approach that does away with a single long program in favour of a series of small self-contained programs. Each self-contained unit or 'object' handles one function and objects are called upon as they are needed. In Cosmos every program is equal to an organism. Within this organism are a series of smaller routines, each one the equivalent of a cell. Each cell handles a separate process. All the cells within an object share the same bit string of instructions but they can execute different parts of it, in the same way that the specialized cells in multicellular organisms carry out different tasks even though they all share the same genes. Taylor has inserted a mechanism that ensures the cells stay dedicated to their task. He has yet to see this specialization emerge but he is planning experiments to try and bring it out. Every cell has the potential to create a new organism. In Tierra there is only one process by which a new organism can be created. This difference means that Cosmos has potentially many more routes to novelty and innovation.

As well as adding instructions Taylor also removed some commands he felt made it too easy for Tierran organisms to evolve. In particular he took out the instructions that allowed Tierran organisms to use the genes of other cells. One of the main features of Tierran evolution is the appearance of many sorts of parasites and hyper-parasites. Taylor thought that this had more to do with this instruction than with the accuracy of Ray's model.

The differences between Tierra and Cosmos do not stop at the level of the organisms each one uses to study evolution. There are significant changes to the environment that Cosmos creatures find themselves in too. The Cosmos world is not just a chunk of memory. Instead, like a cellular automaton, it is set on a 2D grid and each cell occupies a specific position on the grid. Again Taylor felt this was more realistic than Tierra, in which any organism can steal information from any other organism even if it is effectively on the other side of the Tierran world. Cosmos creatures can only spawn offspring into the spaces around them.

Scattered around the grid are energy tokens that the cells have to consume to stay alive. Every time a cell carries out an instruction it has to pay for the privilege by expending tokens. Those cells without enough energy to keep themselves ticking over wither and die. Taylor inserted this condition to make his model more lifelike. 'In Tierra there is something slightly wrong because a program is allowed to run instructions without having to do anything for it,' says Taylor. 'If you look at biology, organisms are having to compete and they have to have energy in order to survive.' Taylor is convinced that this is a key feature of evolution, the life and death struggle for resources. All Tierran organisms fight over is living space.

This innovation also opens up the route to cannibalism and the emergence of predators that feed off other cells for their energy. Using this Taylor hoped to provoke an evolutionary arms race between the organisms, thereby driving evolution and adaptation.

In another important change the cells within Cosmos can communicate between themselves and organisms can also share information. An ability that has only recently been shown to be important.[39]

Although Taylor presented early inconclusive results from Cosmos at a conference in Brighton in July 1997, since then he has spent a lot of time combing the bugs out of the system. 'It is very difficult trying to debug something like this if you have got programs evolving by themselves,' says Taylor. 'It finds all sorts of situations that you did not think of.'

The early runs of Cosmos have shown that it does behave differently to Tierra. In Ray's original system, and to a lesser extent the Network version, organisms with beneficial adaptations spring up and spread throughout the population very quickly. In Cosmos the rate of evolution is much slower, fewer new organisms emerge and the rate of reproduction and turnover in the size of organisms is smaller. Taylor expected this because in Cosmos there are fewer ways for cells to communicate and share genes. In Tierra cells can directly read instructions from other cells. Says Taylor: 'This emphasizes the fact that an important factor governing the behaviour of these systems is the specific design of the language in which the programs are written, and the rules governing how they interact with the environment.'[40]

The stately pace of Cosmos may be more indicative of the gentle jog of evolution in the real world, but Taylor says that it has as much to do with the design of the software. Taylor found that occasionally programs gradually increased in length but when they did so reproduced less often. Taylor repeated a few runs in Cosmos to see if the same thing happened again, thinking that he was perhaps seeing the start of the Cambrian fuse being lit. Unfortunately every time he did the fuse sputtered and died. Although he did see the phenomenon happen again he was forced to conclude that it was the results of lucky accidents and it would take many more accidents to produce as big a bang as that seen in the Cambrian explosion. Taylor is now exploring the effect that accidents can have on the way that a population evolves.

In every other respect Cosmos behaves just like Tierra, populations spring up and die back, new organisms emerge and gradually come to dominate. Even though he tried to make his model more realistic Taylor is left with the feeling that it falls so far short of the animal world that it tells us very little about evolution there.

In one sense this is not a problem. None of the creators of models like Tierra or Cosmos seriously thought that what they were programming was an exact replica of real life. So failing to reproduce every evolutionary leap and stumble recorded in the fossil record does not undermine what they are doing. The value of models like Tierra, Cosmos and Avida lies in the fact that for the first time they allow researchers to compare how similar organisms spring up even though the starting conditions are very different. It lets ALife researchers roll evolution forward and back and helps them understand what happens every time in painstaking detail, something that is impossible with living systems. But now Taylor and others are beginning to question if these models can tell us anything about evolution on a wider scale at all. Perhaps all that has been discovered is the evolution-ary dynamics of a few isolated artificial worlds and nothing about life on Earth.

Taylor is not quite so pessimistic. He believes there is worth in what Ray, himself and Chris Adami are doing. But he believes that the way forward is not through models that are more faithful to the real world but ones that are more abstract. 'The more I have been thinking about self-replicating programs, the more I have been wondering whether the Tierra approach is a good one to take,' he says.

It is likely, though controversial, that evolution started with RNA and DNA and in a series of steps produced new forms, each one of which incorporates a change or improve-ment that the others lack. Stepwise change is easy with DNA because it stands several steps removed from the processes that do the reproducing. Tierra and its imitators take a self-replicating algorithm and try to evolve that stepwise into new forms. Getting from one algorithm to another in a stepwise fashion so as to imitate an increase in adaptability is much more difficult. Taylor suggests that there are many more routes from DNA to a variety of animal forms than there are from algorithm to algorithm. Consequently it is much less probable that a few thousand runs of a computer program like Tierra will produce the same kind of change that has been seen with DNA. This does not mean that trying to simulate evolution is a hopeless task, but it could be a tougher problem than most have thought and it may take a lot longer than anyone thought to produce anything resembling the Cambrian explosion. Taylor says that it may be useful to look at models in which the organisms doing the evolving are not algorithms at all. Instead something more akin to DNA and RNA should be used.

Taylor is now working on another ALife simulator that is very, very different from Tierra, Avida and even Cosmos. He calls it Nidus (the word means nest) but at the time of writing he was still designing it. With Nidus Taylor wants to get away from organisms and cells and move toward something more like artificial chemistry (like Barry McMullin and Walter Fontana). With such a system he hopes to be able to explore more fully the kinds of events that can happen at critical evolutionary boundaries.

Calling All the Ants

Tom Ray is not the only ALife researcher to be inspired by the antics of ants, but the other researchers who take their cues from these insects are not satisfied with just watching them. Instead these folks are trying to create software ants that act just like the real thing. They are not doing this to study how ants go about building nests or foraging for food, instead they have a much grander task in mind. They believe that hordes of stupid ants could be really good at helping manage one of the most complex creations of the modern world, a system that has been called by some commentators 'the largest machine ever built.'[41] The system in question is the international telephone network. Telecommunication companies are very impressed by the potential of software management systems modelled on ants to run their networks for them. This is becoming a pressing concern because the ability of the engineers to keep the phone network going by more conventional means is rapidly diminishing.

The main group working on this approach has not come hot and sweaty from the rainforests of Costa Rica, instead they toil in a much more prosaic location. Ipswich in the UK. Actually the location in question is just outside Ipswich, and is known as Martlesham Heath. It is the home of the research laboratories of British Telecommunications (BT). This windswept and occasionally bleak corner of Suffolk may not be sun-drenched like Xerox Parc in Palo Alto, or have the hippy roots or any number of California campuses, but the ALife workers at Martlesham may ultimately be just as influential.

In 1994 Simon Steward and Stephen Appleby, two software engineers at the labs, authored a paper[42] that suggested that small, independent computer programs that acted like ants might be very good at running telecommunications networks. In passing they noted that humans were getting progressively worse at keeping these networks going.

It used to be easy to run a telephone network. In the early days of telephony when it was a novelty to use a few bits of Bakelite and wire to talk to friends hundreds of kilometres away, people did not use the telephone much. And when they did they did not spend long on it. They made two to three calls per day, each one lasting a few minutes. There were a few peaks in demand but never enough to swamp the capacity of the telephone lines to carry all the calls. The number of calls was governed largely by random events.

Today things are very different. Not only do we have phones at home and at work, we also have mobile phones, fax machines, pagers and mobile computers. People spend a lot more time on the phone than they ever used to, and they don't just use it for talking either. Low local call charges combined with deals that let people use the Internet for as long as they want are encouraging some people to tie up lines for hours. The growth of teleworking means that others may be using the phone line to dial into the office LAN so they can pick up and send e-mail while they are at home. There are radio stations launching phone-in competitions on the spur of the moment. There are special deals offered by airlines that ask people to call a specific number at a certain time. Mobile phone companies in the UK that have been offering free local calls have found that some of their customers use this as a free baby listening service. The parents call their home phone from the mobile and then toss it into the cradle with their child. If the baby wakes or is in distress the parents sitting downstairs will hear it. Whenever there is a crash on a major motorway and traffic gets backed up behind it, one of the first things people do is get on their mobile phones and inform their families or the people they are supposed to be meeting that they are going to be late. The list goes on. Suddenly running a telephone network is not so simple any more; worse, it is completely unpredictable. When someone picks up a phone today neither BT nor any other phone company knows how long it will be before the receiver is put down again. It could be a short call to a pizza delivery company, a chat with a friend or a phone-phreak embarking on an all night war-dialling session to discover which phone numbers connect to computers and are ripe for attack.

The effects of these changing patterns of use are already being felt. Those who insist on using the phone for just talking are finding that sometimes they cannot get through because so many other people are using it for purposes it was not originally designed for. In the US this has already led to some telecommunications companies being unable to ensure that those making ordinary phone calls can get through. Most of the capacity is being sucked up by hungry Net users.

In Europe the story is the same. France Telecom estimates that the amount of Internet data flowing through its network is increasing by around 15 per cent per month. Already the growth in use has forced the company to devote more capacity on its transatlantic lines to Internet traffic than to simple telephony.

Every study conducted on the growth of the Internet suggests that the network is soon going to be used for much more than just the browsing of Web pages. Already Internet telephony and faxing are taking off. In these applications telephone calls that would usually be made over expensive long-distance lines are sent instead over the Internet. All the caller pays for is the local phone rates at either end. Other firms are mooting broadcasting videos over the Web, though it is far from clear whether such services will actually make any money.

All of these new applications pose a serious problem for the telecommunications companies, who are used to gradually growing their networks to meet slowly rising demand.

With the growth of the Internet, demand for capacity grows in leaps and bounds. The telecommunications companies are having to react faster and faster to meet customer demand. While they are just managing to keep ahead of demand for capacity their ability to manage such burgeoning networks is not keeping up.

BT, and most of the other telecommunications companies on the planet, manage their national and international networks using a centralized computer system. In BT's case it takes the form of a mammoth computer program called the Customer Service System (CSS). This program went live in 1989 and in January 1997 was responsible for running the systems that bring in 70 per cent of BT's annual £15 billion income. CSS handles everything to do with billing, fault reporting, fault fixing and orders for new lines. Using CSS, BT operators can connect or disconnect a line for any of its 24 million customers using a few simple typed commands.

Since it was installed the CSS program has been growing at a rate of 22 per cent per annum.[43] As of January 1995 (the last period for which BT made figures available) CSS was handling 28 million exchange connections. It was also producing 433,000 bills and handling 34,000 repairs every day. The program itself is made up of 30 separate systems each one around 500 gigabytes in size. It is made up of millions of lines of code, a distance that the software engineers at BT are proud to claim is measured in kilometres.

Although CSS has been a resounding success it is coming to the end of its useful life. A report about CSS written for BT's *Technical Journal* and published in January 1997 notes simply: 'Its design does not meet today's requirements.' It mentions CSS's inflexibility and notes that this is preventing the company from reacting fast enough to changing market conditions. It gives BT five years to find an alternative. The report notes that 'a big bang solution is not practical even if it were desirable – it would take years, cost millions, and be very high risk'.

The limitations of CSS are becoming clear. One of its key systems is a database containing the most up-to-date record of how all the different parts of the network connect to each other. Changes are made to this database and the network as the load of telephone calls changes. If, for some reason, some of the lines used to pipe telephone calls around the country suddenly go dead because a telephone exchange blows up, the engineers can quickly re-route traffic so as few calls as possible are cut off. By carefully pre-planning what to do if one bit of the network dies BT is rarely caught on the hop when a piece of equipment does fail. Unfortunately for BT's contingency plans, switches are becoming more robust all the time and mechanical failures are rare. Now the biggest problems BT has to contend with are cable breaks and sudden surges in the numbers of telephone calls being made that can threaten to overwhelm a switch or exchange. The problem is not a component in the switch failing but simply the fact that it runs out of lines for the calls passing through it. By their nature these events are hard to plan for. This can mean that when engineers call on a contingency plan to cope with a cable break or radio phone-in the tactic they wanted to employ has already been used by another engineer to cope with a surge somewhere else in the network. The network is changing too fast for CSS, and any other similar system, to cope. Part of the reason that CSS gets more unwieldy every year is because it has to be constantly extended to cope with the ever-changing telecommunications network. As Appleby and Steward put it: 'As systems become increasingly more complex, the weaknesses of central or fully distributed control are harder to overcome. Systems with central control suffer from poor scaling with system size, due to the increase in communication and processing, and are potentially vulnerable to controller failure.'[44]

Peter Cochrane – head of advanced concepts at BT and a man with an infectious enthusiasm for ALife and how it might be able to solve these problems[45] – has one of the best jobs, and job titles, on the planet. He is employed to find out about subjects like ALife and Complexity theory and see how BT can turn the insights of these fields to its advantage. Admittedly the advantage BT is seeking is fairly mundane. All it wants to do is spend less money on software engineering every year. If it can make a tiny difference to its annual bill the company will save itself millions.

This goal of reducing spending on software is not just a nice idea. Cochrane thinks it is essential that BT finds a better way to control its network. If it cannot, then it faces a grim future. CSS is growing at such a rate that soon BT could be spending most of the profits it makes patching up an old program that gets progressively worse at the job it is supposed to be doing. Cochrane says the danger for BT, and by implication any other telecommunications firm that runs its network using something similar to CSS, is that soon it will be spending half of its money and resources on CSS in order to watch what the other half of its resources (the network) is doing.

What BT needs then is a way to replace CSS bit by bit with a distributed system that can react swiftly to changes in demand. Appleby and Steward think the best way to do this is to use ants. The insight behind this thinking is fairly straightforward. The pair realized that although individually ants are pretty stupid, collectively they manage to build nests, regulate the temperature within them, forage and mark food so it can be found by other nest mates. They manage to do this despite the fact that they only have a few hundred thousand neurons and rely largely on scent to find their way around. They note: 'Ants, after millions of years of evolution, seem to have developed strategies that exhibit the properties we would like in our control systems.'[46] The properties in question are robustness and stupidity.

The secret of the ants' success is that they use the world as their memory. Instead of trying to remember landmarks, locations and events ants simply react to what they come across. What they find determines what they do, even if they find something that they, or a sister worker, has changed. Using the world in this way, as its own model, is known as stigmergy. It is a concept that robot makers Rodney Brooks and Mark Tilden are familiar and happy with. What it means for telecommunications managers is no more central-ized out-of-date databases that fruitlessly try to keep up with every event that takes place out on the wires. Appleby and Steward's suggestion is that they should let lots of ants roam around the network and react to what they come across. There is no need to remember where the problems are because, as long as the problem remains, they have to deal with it. Once its gone they can forget it was ever there. The ants should be allowed to manage the network as it is, rather than be told what to do by a database that no longer knows what it is talking about.

In designing the ant-based management system Appleby and Steward wanted to capture the features that have made ants so successful. The ant programs would have to be present in large numbers; be highly mobile; able to leave stationary messages that fade over time; and have the ability to wander randomly. Finally they would have to be simple creations rather than complex programs.

Specifically the Steward–Appleby ant model employs small computer programs, no more than a thousand lines of computer code, that have the ability to wander randomly around the network. To do this they are given the ability to copy themselves from switch to switch. These ants, or agents as the two researchers called them, act like queen ants. Every time they pass through a switch they check to see how it is handling its load of telephone calls. To do this it will consult the information held on the switch that reveals how much spare capacity the switch has available. Most of the time the queen ants will find that all is well so they will have nothing to do. They will then randomly choose a route away from that switch and move on to the next node in the network.

Occasionally, though, the queen ants will arrive at a switch that is struggling to cope with the amount of telephone traffic passing through it. The surge in traffic could be caused by any one of several different events, a cable may have been cut, diverting a lot of extra traffic on to that switch, a radio phone-in might have been announced, anything. The queen ants don't know or need to know why this is happening. All they need to know is that the switch should be running better. At these times the queen ant will migrate to the node that is sending out too much traffic. Once there it will spawn a colony of smaller programs. These worker ants will scatter away from the overloaded switch to all the other nodes that the struggling switch is linked to. At each of these the worker ants will look at the address book or routing table that each node maintains. This table is a list of all the switches that it is connected to as well as the amount of telephone traffic flowing through that connection. The ants update this list and divert traffic on to underused routes. This way the load on the struggling switch is gradually eased.

Appleby and Steward have tested ant-based management systems on a mock-up of BT's core network. Into this they fed a traffic profile that would overload several of the nodes. The system they were testing had a relatively small number of nodes and only two queen ants were needed. Appleby and Steward watched as the queen ants worked out what was happening and launched smaller ants to cope with the load. More worker ants were launched as the amount of traffic on the network grew. Before the overloaded switches failed the worker ants managed to adjust the network configuration until traffic was spread evenly and all the nodes were coping. Once the workers had done their job they were killed off, but the queens kept patrolling and keeping an eye on traffic patterns. The pair concluded that the ant system was 'a powerful way of introducing a degree of intelligence' into the system. Cochrane is excited by the ability of these ants to manage a network despite their simplicity. If BT could be persuaded to use large numbers of ants on its own network it could throw away millions of lines of code it uses to manage the network and concentrate on billing and setting up new services.

Despite the success of Appleby and Steward's work others have suggested that it is unnecessarily complex. They say the same job could be done with even simpler ants. What worries these other researchers is the large number of worker ants that are launched once a queen ant has reached the overloaded switch. Although the worker agents are small, a few hundred bytes in size, launching lots of these programs on an already struggling switch may be the last straw. There is also the, admittedly remote, possibility that a queen agent will become corrupted. Because the queens are relatively powerful, if one goes wrong, it could do a lot of damage before it is stopped.

The other ALife groups working on ant-based management systems are avoiding such problems by making their ants even simpler and even more lifelike. One such system is being developed at the research laboratories of Hewlett-Packard in Bristol by Ruud Schoonderwoerd and Janet Bruten, who are collaborating with Owen Holland from the nearby University of the West of England.

The species of artificial ant that this trio is planning on using is much closer to real ants than those of Appleby and Steward. Schoonderwoerd and his colleagues do away with queen ants and stick with the workers. The different nodes in the network create a constant supply of ant agents and unleash them on to the network. Once launched these ants randomly pick a route to another network node and travel to it. Once they reach that node they die.

On their one and only journey the ants contribute to the management of the network. As they move across the network they leave a trail behind them that is very similar to the pheromone scent trails that real ants use. In the real world these trails guide worker ants to food sources or the fastest route round an obstacle. Every time an ant passes through a switch it leaves a marker behind. In Schoonderwoerd's plan the more traditional routing tables that switches use to keep track of the network are replaced with a table that sums the smells or markers left behind by the ants. The ants adjust the table in the switch that applies to the route they are travelling. The scores record the success and speed with which previous ants have travelled that route. Markers fade over time, so fast routes tend to be well marked and high scoring. When choosing which way to move ants always travel down the routes with the highest scores. Congested nodes will slow down an ant and lower the score of a route. The path will become unattractive to the ants and they will avoid it in favour of higher scoring routes.

This process of leaving and following trails is an exact analogue of what happens in the real world. Living ants spend a lot of time finding nothing as they randomly wander across the forest floor. Occasionally, inevitably, they find something tasty and something that they just have to tell their sisters about. They then mark the food and wend their way back to the nest. On their journey back they leave a series of scent markers to guide other ants. Any sister ants from their own nest they encounter on this return journey are told about the food. These ants follow the scent trail back to the food supply and start ferrying it back to the nest. The scent trail laid by the ant that discovered the food may not be the fastest way back to the nest but that rarely matters. Other ants from the same colony that have either been told about the food, come across it themselves, or have followed a trail to it may forge their own paths back to the nest. In this way it will swiftly become obvious to every ant which is the fastest trail because it will be the smelliest. More ants means more scent, fewer ants taking more time means an increasingly weaker smell, as the pheromones break down over time.[47]

It is this process that Schoonderwoerd, Bruten and Holland have captured with their simple nomadic ants. The researchers have conducted trials of their ant system on the same mock network that Appleby and Steward used. They found that it managed to control traffic loads just as well as the queen and worker ants. It also managed to avoid some of the problems of the Appleby and Steward model, which occasionally manages to set up circular routes. Because Schoonderwoerd's agents are always travelling to a specific destination rather than randomly wandering this is a fault they never fall prey to. If some of the ants are corrupted and stop working properly, the system will neither come to a halt or go catastrophically awry. Instead it will slowly and gracefully degrade, giving the engineers ample time to deal with the problem.

The third group working on ant-based management systems is led by Marco Dorigo and is based at the Free University of Brussels.[48] Working with Gianni di Caro, Dorigo has created a system that is designed for data rather than telephone networks. The problems of routing data are very different to those that telecommunications companies struggle with. On telecommunications networks time is all important. Conversations are digitized and split into data packets for transmission and all the parts of a particular conversation have to arrive within a certain time and in the correct order. We can still make sense of conversations that are slightly scrambled, but if the delay gets too great or the gaps between words or phonemes are too great then we begin to struggle. Telecommunications engineers call this 'predictable latency'. If the time limit is exceeded the conversation will become unintelligible as the bits of data arrive in the wrong order.

On data networks information is split into packets and these are sent across the network willy-nilly. Often they take very different routes and amounts of time to reach the same destination. It makes no difference what route they take. Once all the packets of data have arrived they are reassembled in the correct order.

Part of the reason they travel this way is because of the way that routers work. Individual routers pick a path for a data packet based on the information they hold in a local address book. This address book, or routing table, is regularly updated to take into account network nodes that appear, disappear, or become overloaded. In a busy network the routing table may have to be updated very often to cope with surges in traffic. Making such a network run smoothly is a formidable task. If conditions change too quickly router storms can develop. When this happens the routers spend all their time updating their routing tables and then passing the changes on rather than forwarding data.

Dorigo's ants are launched at regular intervals from routers. They are quite sophisticated ants because they have a memory and a timer. They randomly pick a destination node and set off to travel to it. They use the same routes as the data packets travelling across the network, so if there is any delay on that route they will be slowed down. On the way they time how long it takes them to hop between nodes. When picking which node to travel to the ant looks at the available nodes and picks one at random. It prefers nodes that have not been visited recently by an ant. Once it reaches its destination this first ant despatches another ant that uses the original ant's memory store to trace the same route back. It gets back faster though because it uses a privileged communications channel.

As the ant passes through the nodes back towards the starting point it adjusts the routing table of the node it is passing through depending on how good a route the ant picked. Fast routes quickly become highly recommended. In this way Dorigo's approach avoids some of the problems that the original Appleby and Steward model suffered. Circular routes are rare and there is little danger of ants overloading the network because their numbers are kept small.

This approach of using collections of small smart programs to get a job done is rapidly gaining fans. Typically the programs are called 'agents' rather than 'ants' but the central idea is the same. The main difference between the two approaches is the size of the programs involved. Ants tend to be small and stupid – good at doing one or two tasks but nothing more. Agents are larger and can carry out more general tasks. Pattie Maes at the MIT Media Lab is one of the leading researchers in the agent field.

The agent approach is a radical departure from the traditional computer programming approach. Usually programs are huge monolithic entities that work serially. Everything the computer can do is held in one big program. Agents are smaller, smarter cousins of these larger programs. Large applications like spreadsheets can be built up using collections of agents in an approach that is known as object-oriented programming. This splits programs into functions and creates objects, or agents, to do the separate functions. Crudely when a user of a program wants to do something, such as produce a graph, the objects that handle addition, calculations and display are called up and do what they are told. Instead of each program having to be created individually and able to do everything itself objects are passed around between programs and re-used.

Other agents act alone and do specific tasks for those surfing the Web or searching through a database for some information. In a seminal paper entitled 'Agents that Reduce Work and Information Overload',[49] Maes laid out her vision of how such smart, discrete programs can help us cope with the glut of information that many people have to sift through on a daily basis.

The agents Maes has in mind can learn what you are interested in and filter the news sent to your e-mail box so you only see what interests you. Maes says that the problem at the moment is that users have to do everything them-selves, we are still stuck with the 'direct manipulation' interaction metaphor. Perhaps it would be better to create agents and delegate to them the task of trawling the Web for information we are interested in rather than spending all our time searching for it ourselves.

As an example of what these agents can accomplish Maes has set up a company called Firefly. This Internet-based company offers a range of services based around the use of smart agents. In one of the most popular services visitors to the Firefly web site feed in their music tastes and the agents make recommendations based on the user's profile.

There are problems with the agent approach that remain to be solved. The agents that go out and trawl the Internet for subjects and information have to travel from site to site sampling the information that they find. Unfortunately a lot of Web sites do not allow this kind of access because this is exactly what computer viruses do. No one has yet come up with a way to help a firewall on a Web site distinguish between viruses and agents.

An allied problem is that there is no accepted definition of what an agent is. Consequently agents come in all shapes, sizes and configurations. This causes problems because it means that every agent is essentially a lone operator. It might be much better if it could ask an existing community of agents if they have any of the information it is looking for. The problem is that there are no accepted ways to do this, there is no lingua franca among agents.

Agent-based approaches are springing up in many different places. Many programmers like their flexibility. Making changes to a program usually means creating a new agent and deleting an old one rather than having to make changes to the lines of code. Agent-based control systems are being developed for production lines and even nuclear reactors.[50]

The ant/agent model can be extended in many ways. BT also has plans for taking the idea much further than Appleby and Steward's original model. In fact it wants to take the ant system even further than Dorigo or anyone else has before and make them as lifelike as it gets. BT wants to make the ants breed.

Chapter Five

Living Machines

Someday a human being may shoot a robot which has come out of a General Electrics factory, and to his surprise see it weep and bleed. And the dying robot may shoot back and, to its surprise, see a wisp of grey smoke arise from the electric pump that it supposed was the human's beating heart. It would be rather a great moment of truth for both of them.

Philip K. Dick, *The Android and the Human*[1]

Good for Nothing

Sex is a problem. It makes no sense. In fact, sometimes you wonder why we bother. But sex is not a problem just because it causes innumerable arguments amongst us humans, it is a problem because biologists are still not entirely sure what it is for.

Swapping genetic material has been around for a long time. Simple bacteria discovered it over three billion years ago and have been indulging in it ever since. Bacteria do not have sexes. They may happily swap genetic information between each other but they reproduce asexually, by budding. There is no split into separate mating types, no bucks and does, no cobs and pens, no males and females.

The rest of us are not so lucky, we indulge in a different sort of sex – a kind that demands a mate – and this is where the first problem with sexual reproduction rears its ugly head. Often animals have to go to a lot of trouble to find a mate, time that might be better spent reproducing by themselves. As renowned biologist John Maynard Smith, who has spent more time than most studying sex, puts it: 'It is not merely that sex seems pointless; it is actually costly.'[2] His point is that if I could mate with anyone then I have a lot more potential partners than I do if I can only mate with half of the members of my species. Logic suggests then that sexual reproduction should be rare. Unfortunately it is everywhere. There are some species that can clone themselves and reproduce asexually, such as dandelions,[3] there are many more species that do have different mating types.

Another, related, sexual conundrum is the fact that most of the species that do mate have only two mating types. ALife researchers at BT have studied the dynamics of populations with more than one sex and gleaned some insights into why two is the most common number. But more of this later.

Given that it takes a lot of effort to find a mate, the question is why organisms persist with it. It is certainly an idea that has caught on. But no one is sure why they keep on having sex when there are such advantages to being asexual. Maynard Smith posed the problem in terms of what would happen if suddenly an organism that previously had reproduced by mating with the opposite sex forswore heterosexual relations and decided to go it alone. If they relied only on their own genetic material to reproduce, the clones they produced would swiftly come to dominate. Given the overwhelming advantage that would accrue to any organism that gave up sex why do so many species, like incorrigible sinners, still keep on having sex?

Sexually Combated Disease

For many years it was thought that what made sexual reproduction so persistent was the advantage it gave those organisms in dealing with environmental change. Sexual reproduction with its swapping of genes is the only way to ensure a constant supply of offspring that can cope with the challenges that the environment is throwing at a species. In consequence you would expect that sex would be common among species that are constantly under threat in environments that are fiercely contested and rapidly changing. Unfortunately this is not the case, the opposite is true. Sex is most common among large animals that live a long time and have few young.

This makes the persistence of sexual reproduction even more puzzling. Most of the populations of these animals are well established, well fitted to the environment they are part of and they have little to compete with, bar the odd predator. There seems to be no reason for producing young that are slight improvements on older models. They seem to be using reproduction to keep competing but there does not seem to be much to compete with. Why do they keep doing it this way?

For a lot of reasons sex is a problem. Explanations have come and gone but none seems to fully explain what sex is for. Biologist Matt Ridley summarized it well when he wrote: 'We are left with an enigma. Sex serves the species, but at the expense of the individual. Individuals could abandon sex and rapidly outcompete their sexual rivals in passing on genes. But they do not.'[3] The question is: Why not?

The answer is disease. Organisms need sex in order to fight off infections. Without sex they would have no way of coping with the endlessly creative ways that diseases regularly invent to attack cells.

Most diseases work by subverting cells. Over the eons disease organisms have become very adept at using other cells for their own ends. While fungi and bacteria tend to consume cells, viruses break into a cell and make all the machinery inside work for it rather than for the host.

Usually a cell stays closed to outsiders thanks to proteins on its surface that act like locks. Unlocking a cell and getting inside requires a key in the form of a protein of the correct shape. Disease organisms are really good at finding out which protein unlocks a cell. So what the body needs is a way to change the locks once viruses and bacteria have got hold of the keys. Sex is that way.

The longest arms race in history, one that stretches back to the dawn of life, is the struggle between organisms. The spoils of the fight are the keys to the kingdom of cellular activity. Once a disease organism has found a key it will spread swiftly through all the other organisms whose cells can be unlocked with that key. Sex helps organisms change the locks on a regular basis and keep the parasites out. The different locks are made by separate versions of the same gene. Usually these genes are spread throughout a population. Sex helps the versions get around and as a result helps a species fight off disease, keep its cells intact and stay healthy.

This idea that sex helps organisms fight off disease goes by the name of the Red Queen Hypothesis. The name comes from the frenetic living chess piece that Alice encounters in Lewis Carroll's *Through the Looking Glass*. Although the Red Queen runs as fast as she can she never gets anywhere, the best that she manages is to stay in the same place. The same is true of sex. It may not help an organism get ahead but it ensures that the parasites chasing it have to run very hard to keep up.

In the late 1970s this theory was tested using an ALife model. Oxford biologist Bill Hamilton built a computer model that was populated with 200 creatures. Some of the creatures were sexual and the others were asexual. Death was randomly visited on the creatures and as logic suggests the asexual race quickly came to dominate. Next Hamilton introduced parasites whose ability to infect and weaken the creatures depended on the virulence genes they possessed. The original creatures fought off the disease using resistance genes. When the simulation was run again the asexual population no longer had an advantage and the sexual species often came out the winner. Some biologists have criticized Hamilton's model for its over-reliance on parasites and its inability to explain the ubiquity of disease, but others believe its conclusions are worth serious consideration.

Phone Sex

But what has sex got to do with BT and telecommunications networks? Although biologists are coming round to the view that sex is a method by which organisms renew their defences against disease, it also remains the route by which some innovations, novelties and adaptations are introduced into populations. There is no doubt that mutations in DNA or errors when genes are copied can profoundly affect how an organism copes with what the world throws at it. The same might be true of computer programs that can mutate, change and evolve. Making computer programs breed might be a great way to help them cope with a changing telephone network. This is how Peter Cochrane and his colleague Chris Winter at BT want to use sex, to improve their computer programs and make them evolve.

Cochrane and Winter are conscious that although the Appleby and Steward ant system is very good at managing the internals of a telecommunications network it has a serious drawback. The ants in it are static, they never change. Once the job of the worker ants is done they die off. The 1000-line program that is an ant will not change. In some respects there is no need for it to do so. As long as the ants are given relatively low-level tasks such as managing the telephone traffic on the nodes in a network, there is little need for them to change. This is in contrast to the telephone network, which is in a constant state of flux. What Winter and Cochrane would like to do is create ants that can evolve and adapt to the changing conditions of the network. Ideally they would like to seed the network with ants that can breed and let evolution create a world-beating network management agent.

There is an argument for saying that ants are the fruit of a 90 million-year engineering project and the design is as good as it is going to be. But the same is not true of the ants that Cochrane and Winter want to use. These capture only a small sub-set of any behaviour and may react badly to the swift changes in the way that the telephone network is organized. They need a way to develop new capabilities to handle situations they have never encountered before.

Certainly the uses that telecommunications companies make of ants will not stop with the Appleby and Steward system. Cochrane says that US telecommunications company MCI is already using the ant-based system. In doing so it replaced millions of lines of code with a few thousand and saved itself a lot of time and money. The project to use the ants was carried out because MCI and BT were planning to merge their businesses. Because of the problems involved in merging telephone networks, each of which serves 20 million customers, both companies were interested in looking into new ways to manage their huge networks.

Unfortunately the merger deal between BT and MCI has broken down and it is now highly unlikely that MCI will be the one pushing ahead with BT's ALife agenda. To make matters worse BT cannot go it alone and implement the system on its own telephone switches because they are too difficult to modify. The digital System X switches used extensively throughout the UK telephone network are incapable of supporting distributed programs like the ant system.

This setback aside, Cochrane and Winter are pushing

ahead with ALife. They think that making computer programs breed could be the way to tap the innovative and creative force otherwise known as evolution.

Sexual Searching

Babies, puppies, kittens, larvae, call them what you will, but all offspring have one thing in common. Whether they have brown eyes, feathers or scales, they all represent a single point within a space of possibilities. A goodly proportion of the total range of possibilities available to a species will be found in the living population of a species be they people, dogs, cats, or flies. The entire range of possibilities is known as the gene pool or genome.

There are many ways that children or offspring can differ, be it through hair colour, eye colour, nose shape, or height. These properties can be made to represent the axes of a map – that in many ways resembles a phase space. The population of children, or kittens, or goslings can be plotted on this multi-dimensional graph. Each kitten or child has its position defined by whether it has blonde hair, blue eyes or a ginger coat. Plot enough points and you start to get a landscape of possibilities emerging – you start to find out about how deep the gene pool for that population is. The landscape is unlikely to be smooth either, it may be crossed by chasms or fringed by mountains representing those qualities that are rare or common in a population. Before a puppy or child is born there is no way of knowing where it will fall on the map, no means to find out about its abilities, no way to tell whether it will fall into a ravine, scale a mountain or plant a flag on new territory. All its parents know is that it will fall somewhere within the range of possibilities, somewhere on the map.

It would be hard, if not impossible, to create such a graph for a population of children, especially if controversial qualities such as IQ were included. It would be easier to do for kittens and larvae because there are fewer parameters. It is easiest of all to do with computer programs, particularly if the axes do not represent eye colour but instead capture how well the program performs a given task.

Every point on the landscape represents a possible child, kitten, or computer program. Having children is a way of searching the landscape and producing a possible solution to the problems the world keeps putting in the way of your species. Parents of teenagers may be forgiven for disagreeing with this conclusion. While it would be cruel to breed children or kittens for a particular task and then kill them off if it emerged that they did not possess the qualities you valued, the same is not true of computer programs. All that you need is an efficient and fast way of searching through the possibilities. A way to travel the landscape and find the fittest program. This is where genetic algorithms and genetic programming come in because they are really good at searching such a landscape for the best solution to a problem.

Going Dutch

Genetic algorithms were invented by John Henry Holland, one of the grand old men of computer science. Holland was born in 1929 in Indiana and as a child showed a precocious aptitude for physics and mathematics. His ability won him a scholarship to MIT in 1946 just as work was beginning on one of the world's first electronic computers. Under contract from the Department of Defense MIT researchers were trying to create 'Whirlwind', a machine that would be used to plot the flight of incoming missiles.

Holland got involved with Project Whirlwind and soon became one of the world's first experts in computer programming. His expertise attracted the attention of IBM, which asked him to join a team of engineers working on the logic design for its commercial calculator, a device known as the 701. Holland had distinguished company on the project, working with AI pioneers John McCarthy and Arthur Samuels. The former was to be involved in AI from its earliest days and Samuels was then working on a computer program that would learn draughts (checkers in the US) by playing opponents rather than just following the rules encoded in its program.

Despite the august company, programming the 701 was a frustrating business. Design work was done during the day and through the night McCarthy, Samuels and Holland tested their work by running a simulation of a neural network. What made it frustrating was the fact that the 701 was tricky to program and broke down every thirty minutes. They got it working after a fashion; even better, they showed that their basic neural network could be made to learn something about a simulated maze. This was a lesson that Holland never forgot, that it was possible to make a machine adapt to a problem it was presented with.[4]

At the time Holland was working at MIT, the institution was dominated by Norbert Wiener and the cybernetics movement. There was little room for Holland to explore his interest in making machines learn and adapt. So Holland moved on to Michigan, where Arthur Burks was setting up the Logic of Computers group. Holland could not have picked a better place to continue his work, even though his original reasons for going there had more to do with Michigan's football team than the research being done in the engineering block.[5]

Burks became the editor of John von Neumann's papers when the great man died, but even before he took up this task Burks was interested in cellular automata, neural nets and the links between computers and biology. Holland fitted in perfectly. While pursuing his interests Holland came across a book entitled *The Genetical Theory of Natural Selection*. It was written in 1930 by evolutionary biologist R.A. Fisher and was the first attempt to create a mathematical theory of evolution. The book made Holland realize what a powerful force evolution was with its potential for generating novel and workable solutions to a problem. The key to its success was the gradual accumulation of change, 'a progressive modification of some structure or structures'.[6]

Holland invented a way of tapping this vein of creativity and slowly building up adaptations to find the best way to do something. He laid out his ideas in a book called *Adaptation in Natural and Artificial Systems*. In the book he expounded his thoughts about genetic algorithms. An algorithm is simply a way of doing things. Algorithms come in many forms but are most usually a set of rules. As you would expect from anything called a genetic algorithm (GA), these are rules that change in much the same way that real genes do.

GAs take the form of short computer programs – essentially a string of bits. The different bits in the string are used to encode the variables the problem is working with or the functions the program is being used to combine. The bit strings act as the equivalent of genes in animals. In the same sense they define what the organism or algorithm carrying that gene can do.

Initially a large pool of bit strings is created. All the strings are the same length and use the same encoded variables but the bits are randomly scrambled within each string. The bit strings are then tested to see how well they complete the problem the GA is being applied to. They are scored on their performance. This score is related to a predefined fitness function determined by how well they did their job. Each GA represents a point on a landscape and a search through the space of possibilities.

Once the first run-through has been completed breeding starts. The GAs that did not perform well enough are killed off. A new batch is bred from the remaining bit strings. New combinations of bit strings are produced in two ways – by crossover and by mutation.

Crossover involves sections of successful GAs being combined to create a new GA. This offspring will have some of the qualities of its parents and may end up performing better than they did. Mutations are brought about by randomly flipping bits within strings. Breeding continues until there are the same number of parents as children. Then all the parents are killed off.

Both methods of breeding and improving the performance of the GA represent different ways of getting around the search space. Crossover is the equivalent of a gentle stroll around the landscape based on what is already known. It uses the trails that have already been blazed but pushes them on a little further. Mutation represents radical changes in direction and can bring about dramatic improvements in performance.

Using crossover is a very powerful way of searching. Holland has shown mathematically that using crossover makes searches go much faster than the number of strings in the population might suggest. This is because each string represents more than just itself, it also represents a whole related family of strings. The fitness of each string is the sum of its history. In effect this means that all these related strings are searched every time the GA is tried out. As a result the GA quickly homes in on the optimal solution.

Another useful feature of crossover is that it ensures that chunks of strings are moved around at the same time. This means that groups of genes that work well together usually stay together. Again this means that each new generation is not just a search of individual genes it is also a trial of genes that have become associated.

Travelling man

It is this ability to generate a high-performance program very quickly that BT finds attractive. There are some problems that GAs seem made for because they can produce answers far quicker than any other method.

The problems that GAs are particularly good at solving are known as NP or hard problems. They are hard problems to solve because of the number of possible solutions to them. The NP stands for non-polynomial and refers to the amount of time they typically take to solve. Mathematical problems that can be solved in polynomial time are usually quite easy. The amount of time they take to solve grows by a fixed amount – usually a 'polynomial' function of the size of the problem. With non-polynomial problems the amount of time grows either exponentially or factorially. The archetypal NP problem is known as the travelling salesman problem. This involves working out the shortest route a salesman can take to visit a set number of towns only once. The possible number to check grows very quickly. With five towns to visit there are 120 (1*2*3*4*5) possible routes, but with twenty-five towns, finding the shortest route is the equivalent of finding one raindrop in all the world's oceans.[7]

Although the travelling salesman problem sounds very specific it represents a broad class of problems that businesses regularly find themselves up against. Airlines have to find the best way to schedule their planes so that everyone gets to where they need to be as quickly as possible. Manufacturers often have to work out how all the different components of their products should be put together to speed up production lines. Phone companies have to work out the best way to organize a network so all the calls get put through. The list of situations that the problem applies to is endless.

Luckily GAs are very good at searching through all the possibilities in those metaphorical landscapes. They are faster than any brute force run-through of all the possibilities, especially if you are willing to give up total accuracy for speed. Shara Amin and colleagues at BT have been using GAs to evolve fast solutions to travelling salesmen problems and found that when they ditched the requirement for perfection they got useful results much quicker. Amin found that if they are willing to accept solutions that fall within the top few per cent, they can get solutions to a 100-city problem in seconds, 3000-city tours take around half an hour.[8] Accepting only the best solution would add hours on to the search time. Time that telecommunications companies can ill afford when they are competing in terms of service levels.

What companies like BT want is results that are good enough rather than perfect results. In this respect the ability of GAs to do just this could mean that the ants are soon using algorithms to adapt to conditions in the network. The problem with the telephone network is that conditions change very quickly. Ants that are designed to cope with one situation may struggle when traffic gets very heavy.

If ants have problems coping with novel situations, this problem is only going to get worse if BT moves ahead with its plan to use ants to do more on its network. When different sorts of ants are doing everything from establishing connections to routing telephone traffic and generating phone bills, making sure the ants can cope becomes a real worry. One way to ensure that they can at least adapt is to give them the ability to evolve. Turning them into GAs might be one way to do this. If the ants are given the ability to evolve, there is the problem of controlling this change, however. If the ants are left to evolve willy-nilly, they are unlikely to develop exactly as BT, or anyone else, wants them to. The fitness function of a GA is one way to control the changes that are preserved but it brings it own limitations. If the fitness function is defined too tightly then the programs produced may not be very innovative. The population may quickly stagnate. On the other hand if the GAs are given too free a rein, the programs produced may not do what BT wants.

Dorigo and Schoonderwoerd see no need to evolve their ant programs. Schoonderwoerd says that his ants were designed to do a very specific job and because they do it very well there is no need to evolve them further. Dorigo says that the only part of his ants that might be worth evolving are the rules they use to direct traffic. These rules dictate the tactics that a router should use when trying to pass data on to other nodes in a network. One algorithm is called OSPF, which stands for Open Shortest Path First. Any router using this algorithm will try and direct data down the shortest path to its destination before it tries any other tactic. Dorigo says there may be some benefit in trying to evolve more complicated routing algorithms in a bid to speed up the flow of traffic.

Schoonderwoerd says that making the ants evolve may create more problems than it solves. The problem is that GAs do not scale very well. They are very good for problems that do not have too many variables but once they get beyond a certain point they start to slow down. He adds that it can also be very difficult to define a fitness function that meets the local and global demands of a telephone network. What works for a small collection of switches may not be ideal for the rest of the network. Measuring fitness can be a real problem for creators of GAs.

A final problem with GAs is that they are not very creative. They can innovate within the possibilities defined by their parameters and find a good way to do something but they cannot find entirely new ways to do the same task. How they are defined prevents them from discovering solutions that are not located somewhere in the space they are searching. They are only creative within limits and cannot evolve and change in an open-ended fashion like life can.

Suffer the Children

BT researchers have found that searching goes even faster when they turn the reproductive scheme of the GA on its head. In a typical GA it is the children that survive to be tested rather than the parents. Once a new population of GAs has been produced using chunks of bits from the parents, all the parents are done away with and the new batch is tried out on the problem.

But Jose-Luis Fernández-Villacañas and Shara Amin have found that searching goes much faster if you kill the kids and let the parents get on with it. In the 'virtual children' algorithm that the two developed this is exactly what happens. Just like ordinary GAs they start with a random population of algorithms. Each one of these seed algorithms gives birth to two virtual children. The children are then tried out on the problem the GA is being used to solve. How well they perform is noted by the parents.

The children are then killed off. All the parents retain from these children is information on what helped them do well or badly. Using this information they produce two more children and try them out on the same problem. The children are scored on how they do and then killed off again. The cycle repeats until the optimal solution is found.

Fernández-Villacañas and Amin have found that the algorithm performs as well if not better than GAs on optimization problems. In nearly every search situation they found that it was many times faster than a GA, only occasionally was the algorithm tripped up by the difficulty of the search space it was moving around.

The experience with the 'virtual children' algorithm has

taught the folks at BT a valuable lesson. There are dangers in copying biology too closely and sometimes there are real advantages in improving on biology and doing what it cannot. Software is far more flexible than the wetware that life is. This is certainly the case with the 'virtual children' algorithm and now BT ALife researchers are experimenting with virtual organisms that have different numbers of sexes. They want to see what advantage there is to having only two sexes. Perhaps their simulations would run faster and produce solutions quicker if lots of sexes were involved.

The Flowers and the Bees

Chris Winter and his ALife group have been experimenting with sex in other situations too; in the same vein they have been trying out sexual situations that are rarely found in nature. They have produced groups of artificial creatures that have different numbers of sexes. So far the only use they have put this work to is to help them model business scenarios, but still they have learned some valuable lessons.

Winter, along with BT researcher Paul Coker, has built a simulated world containing seven species. There were a hundred individuals each of six similar species and more than 6000 individuals of a seventh species.

In this artificial world the six similar species are analogous to flowers. To succeed and grow they have to attract the seventh species, an insect analogue, and be 'pollinated'. At the same time they have to expend as little energy as possible on producing nectar to attract the seventh species. All the 'insects' are trying to do in this simulation is gather as much nectar as possible with the minimum outlay of energy.

Although the 'insects' were the same species they were not all identical. In fact they were quite fussy and liked different sorts of features such as colour and scent in the 'flowers'. The 'insects' can easily fly to other 'flowers' to get a better deal. The flower species either reproduced asexually or required between two and six sexes to reproduce.

Winter says that the group assumed that two sexes would begin to dominate quickly because it is so prevalent in the natural world. They assumed that the energy needed to find one mate would only increase if more mates had to be found. They assumed that there must be good reasons for the two-sex system that would become apparent in the simulation. But what they found was that two-sex systems were not always the best. In fact the four-sex flower species came to dominate most often. In certain circumstances six sexes seems to be the best number to have.

What makes the difference is the time that the organism has to react. Winter and his colleagues found that when there is plenty of time and the optimum solution is being sought then two sexes is the best number to have. It produces solutions slowly, but the ones it does produce tend to outperform those produced any other way. When time is short larger numbers of sexes produce good results more quickly, but these solutions are never as good as the best that can be produced by organisms that have only two sexes.

In the real world there is lots and lots of time and this is perhaps why species with two sexes dominate and there is enough room for really good solutions to appear. In the fast-moving world of telecommunications time is a luxury companies cannot afford, especially when they are no longer competing on price and have to distinguish themselves by their standards of customer service and their range of products. This may mean that any ALife agents used on telecommunication networks will need lots of sexes to be able to work fast enough, but the solutions they find will never be as good as the solutions that could be produced if more time was available. Even if the solutions these algorithms produce are not perfect they will still be much better, and react faster, than the human-made versions. We may have to get used to the quirky performance of the telephone network rather than have it working nearly perfectly. Given that the alternative might be no telephone service at all, that might not be such a bad thing.

Winter says that although BT chose the bees and flowers analogy the simulation has applications to the real world. The situation could represent the way that companies work to attract customers by varying their services or even the way that algorithms adapt to control the flow of traffic across a telephone network.

Call for Help

Using artificial creatures to manage your telephone network has its dangers. Part of the reason that ants or agents have not been used on telephone networks is because of their resemblance to computer viruses. Ants and agents wander around telephone networks and have the power to make changes to the switches and nodes that shuffle the telephone calls around. Computer viruses do exactly the same thing, the only difference being that they do it on computers rather than telephone switches. It would be a brave telephone company that unleashed a horde of agents on to its network to do call handling, especially if these agents had the ability to learn and evolve. As Tom Ray has pointed out, 'Evolution is an extremely selfish process.'[9] If these agents are given the ability to reproduce and change, it is likely that any adaptations will serve the survival chances of the species rather than the much more nebulous business aims of telecommunications firms. 'Imagine however, the problems that could arise if evolving digital organisms were to colonize the computers connected to the major networks. They could spread across the network like the infamous Internet worm. When we attempted to stop them, they could evolve mechanisms to escape from our attacks. It might conceivably be very difficult to eliminate them.'[10] Ray was so worried about this that he created a virtual game reserve for the creatures in Tierra.

It may be possible to stop this and drive evolution down certain avenues by careful choice of fitness functions. This may ensure that the ants serve some of our ends rather than their own. But the essence of evolution is its novelty and ability to innovate. This may be curbed if the fitness functions are too tightly defined and an organism does not have the freedom to explore every possibility.

Containing the creatures is another problem. Telecommunications is all about connecting people and it is highly unlikely that any agents released on to one network would stay within the bounds of that network. They may well be attracted to other networks for the very simple reason that life might be simpler there. If there were no other agents on these networks to offer competition, they would be a perfect home. The digital version of the Cambrian explosion may happen when the first telecommunications company unleashes agents on to its network only to see them propagate and set up home on the telephone networks of all the other telecoms companies they are connected to.

If one company is brave enough to take the first step and start using agents, it may prove impossible to recover the network once it has been run by agents for any length of time. BT and all other telecommunications companies have a hard time keeping up to date with the changes on the network when they have all the resources in place to monitor and manage it. If agents look after it for a couple of weeks then crash irrevocably, the company would have a hard time recovering the network and unravelling all the changes that the agents have made.

Get With the Program

BT is so interested in ALife research because of the potential it has to save it a lot of time and money. One of BT's biggest costs is the amount of money it has to spend keeping the CSS software and all its allied programs up to date. Part of the reason for this is the way that CSS has grown in the ten years or so that it has been in service. As CSS grows it becomes increasingly brittle and susceptible to crashing because of conflicting instructions buried within its miles of code. Winter describes software engineering as the last of the handicrafts, because every program is hand made and tuned to a specific job. This makes them beautiful but fragile works of art.

But Winter also thinks that ALife programming techniques can help on this side of the fence too. BT's ALife group is also starting to experiment with the genetic programming techniques developed by another Michigan graduate, John Koza.

The big problem with genetic algorithms is that the length of the bit strings making up the algorithm are fixed and never change. As a result they are usually developed to do one job only. Problems emerge when the string length is increased and the GA is asked to find solutions to a collection of problems. The time it takes to evolve a usable solution increases significantly. This means that it is virtually impossible to use GAs for large-scale programming tasks.

While working at MIT Koza developed an alternative to GAs that produces workable programs and overcomes their scaling problems. In contrast to GAs these genetic programs do not use bit strings, they use nodes and leaves. The nodes represent predefined functions, the syntactical operators that software engineers use to join statements when writing a program. The leaves are the terms such as random numbers, constants and variables that the program is manipulating.

Although the elements within a genetic program differ from those in a GA, the process of evolution is very similar. Initially a random collection of programs made up of collections of nodes and leaves is prepared. These random programs are tested against the problem and scored on how well they do the job. Two of the randomly generated population are then picked with a probability based on their fitness. One child is then produced using crossover techniques with a branch of one tree being grafted onto the other and vice versa. The process of crossover and child production continues until a useful program is produced. Theoretically there is no upper limit to the size of the programs that the genetic program approach can tackle. However, the more complex the problem being tackled the longer it will take to evolve a workable program. This has limited most applications of genetic programming to producing short, compact programs that can tackle problems like pattern recognition. Something that standard programs typically do not perform very well.

This ability to produce short workable programs quickly is what attracts BT to this approach. Winter estimates that within ten years computers will be writing workable programs as fast, if not faster, than humans can.

Software Becomes You

Like its name implies software is easy to chop and change, GAs were originally written to take advantage of this flexibility. For a long time the same has not been true of hardware. Once a chip is fabricated its configuration is set. Any tinkering will just turn the chip into an expensive drinks coaster. Now even hardware is getting softer. Using GAs some researchers are creating the first examples of evolvable hardware – computers that can change and learn.

The increase in flexibility is largely thanks to the development of chips known as Field Programmable Gate Arrays (FPGAs). These microprocessors are effectively blank chips just waiting to be configured. FPGAs were invented by US company Xilinx. Chips from Intel, AMD and National Semiconductor are usually created for one specific task or, in the case of Pentium chips, are able to carry out lots of very general functions that can be harnessed by programs.

In contrast FPGAs are uncommitted, their configuration program changes them into either general purpose chips or single-minded processors that do only one job.

FPGAs are studded with arrays of logic gates arranged in cells, but none of the gates are interconnected, nor is the configuration of the gates set. Turning the chip into something useful requires the loading of a configuration program. Different programs will turn the blank chip into one of any number of different devices.

For some people this is not enough. They want to go further. People such as Adrian Thompson at the Centre for Computational Neuroscience (CCNR) at the University of Sussex, for instance. He is exploring how evolution, via GAs, can create useful circuits on FPGAs. Thompson is your average scruffy, amiable academic and at the CCNR he shares an office with an eclectic mix of robot makers, hardware hackers and animal behaviourists, as well as a nest of wood ants. David Nicholson, another CCNR researcher, is studying the ant's visual system to help create better vision systems for robots.

Usually the transistors on a computer chip are used like switches, being either on or off, so they can channel the 1s and 0s used in streams of data. In fact transistors are capable of doing much more than just switching currents on and off. They can support a wide range of voltages and in some configurations can act like amplifiers. Usually no use is made of these intermediate properties. It is much easier to design a working processor when you only have to worry about a transistor being in one of two states and switching a voltage on and off. Anything else would make the job of designing a chip impossibly complex. Too complex for humans perhaps, but not for evolution.

Thompson is interested in what evolution can do with a raw, unconstrained FPGA. He does not want to force the transistors into pre-defined roles, but wants evolution to discover what resources an FPGA has and exploit them without him telling evolution what to do.

The problem he set evolution was to produce a configuration program that could discriminate between two tones, one at 1 kilohertz and one at 10 kilohertz. Ideally he wanted the output from the circuit to be 5 volts when one tone is played and 0 volts when the other is played. He picked this because, 'It was the simplest thing that I could think to do at the time,' he says. 'I also did it because people said it would never work.' A tone discriminator is also the first step along the way to making a pattern recognizer or signal processing chip – jobs that Thompson has a hunch these chips will be very good at.

For your average electronic engineer able to use any components they want making a simple tone discriminator is a trivial or 'toy' task. The first component they would pick would be an off-chip clock. This would keep the operation of all the transistors on the FPGA synchronized and it would help spot which tone it was listening to. Between ticks the circuit could count the number of peaks in the signal. More peaks means a higher frequency and identifies the 10 kilohertz tone. Even if the engineers were told they could only use transistors, they would probably use them to construct something like a clock. If you stop them using any kind of clock the task stops being trivial and becomes tricky. If you restrict them further and let them only use a hundred cells some would say it cannot be done. 'Many people thought it would be impossible,' he says. Even if they managed to do it, he estimates they would have to use far larger parts of the chip; 'Ten to a hundred times more,' he says.

But Thompson has done the seemingly impossible. Or rather evolution has. It has produced a working tone discriminator that uses only thirty-two cells. and Thompson has no idea how it works.

To give evolution something to work with Thompson first created fifty random bit strings of 1800 bits each. Using these evolution would create the configuration program for the chip. This program would tell the chip how to arrange connections between different cells in the 10×10 corner of an FPGA. One by one the random programs were downloaded on to the chip. The tones were then played and Thompson looked at the outputs the configuration produced. He was looking for the most promising program, the one that took the first step towards creating a tone discriminator. The program he picked produced a 5-volt output whether it was played the 1 or 10 kilohertz tone.

Although this first program was relatively easy to pick, subsequent programs that improved on this can be hard to spot. Thompson says one of the most difficult parts of designing a GA lies in deciding whether a program is closer or further away from the goal. For such a simple program the fitness criteria were relatively easy to work out. Thompson looked for the program that maximized any difference in output when the two tones were played. This meant that the programs evolved in small steps. Thompson adopted this approach because he had one false start with a fitness function that was too selective. All the programs are scored on how good they are at completing the task in hand. The top performers are kept and successive generations of these are mutated and tried out on the chip.

Thompson kept the programs mutating even as they got better at discriminating between the tones. Previous work by Thompson's colleague Inman Harvey has found that once a population begins to converge it can be difficult to drive any new adaptation out of it. To keep evolution moving means throwing the odd spanner into the reproduction process to keep the programs producing novel solutions. Thompson says: 'Sometimes evolution has to be led by the nose and you do that by starting with simple things.'

It took 2800 generations before he was close to getting what he wanted. By this time the output from the FPGA had flipped, however, with high output for the 10kHz tone and low for the 1kHz tone. By generation 3500 the perfect behaviour was produced, but Thompson let it run for another 1500 generations to ensure there was no change.

Once Thompson had his tone discriminator he took a look at the circuit that the program had created. He pruned away all the cells that connected only to themselves and had no effect on the ability of the FPGA to tell the two tones apart. He was left with a circuit that connected together a mere thirty-two cells, far fewer than Thompson thought possible. He tried to figure out how the circuit worked but really he has no idea. One part of the circuit was particularly intriguing. A five-cell cluster had to be kept working to ensure the tone discriminator did its job, even though none of the five cells were connected to the rest of the circuit. Thompson can only speculate about the role that these cells play. Perhaps they were acting as capacitors, but really he does not know. What's more he will never be able to find out. 'They are going to be relying on properties I can't even measure,' he says.

Evolution has been undeniably creative. It has found ways

to use FPGAs that no one imagined. As Thompson puts it: 'Evolution has been free to explore the full repertoire of behaviours available from the silicon resources provided, even being able to exploit the subtle interactions between adjacent components that are not directly connected.'[11]

Despite the success of Thompson's first try at evolvable hardware problems remain to be solved. The program produced by the GA is very tightly tied to the characteristics of one corner of the FPGA. So much so that when it is moved to another corner of the same FPGA the program does not perform as well. The same happened when he evolved a circuit to control a small robot. The control program was evolved on a discrete FPGA on a workbench and was then downloaded into the gate array on the robot itself. The program was supposed to make the robot start moving and stop again when told to. Unfortunately the program did not work at all. To make matters worse the original program works best when the temperature around it is within 10 degrees Celsius of the conditions it experienced while it was being created. In comparison, more traditionally made computer chips keep functioning over a 100-degree range.

To overcome this problem Thompson is starting to create more evolved circuits in batches of five. Each one will be evolved at a different temperature and the worst performing program of the five will be used to seed the next generation. In this way Thompson hopes to start making his programs more reliable, essential if they are to find their way out of the lab and into computers and other domestic devices.

Computer Life

Thompson is in the vanguard of an ALife movement that is working to create evolvable hardware. Hardware that is not hardwired but is flexible and can change and adapt. Clearly the FPGA is the key development that has allowed this approach to emerge. The flexibility of these chips means that the design of a computer chip is no longer fixed, it can change, adapt and learn. Many of the researchers within this field are still using software, usually genetic algorithms, to change the configuration of the chip, but eventually they hope to hardwire everything on the chip itself and remove this reliance on software. The line dividing hardware and software will have been removed entirely, leaving them with something much looser and easier to use. They may be the first to realize von Neumann's dream of creating a reproducing machine.

Daniel Mange and colleagues at the Federal School of Technology in Lausanne in Switzerland are the leading researchers in this field. Some of the researchers within Mange's group helped Thompson develop his chips. Mange dubs his approach Embryonics[12] – reflecting the fact that the work is part embryology and part electronics. The computers the group are starting to create are self-configuring and self-repairing. Given the right instructions and enough blank cells they will also be able to reproduce.

The Embryonics approach begins with an FPGA in which none of the individual cells have been configured. These cells are taken to represent the individual cells of a larger organism. Each one has a small hunk of RAM, a few logical elements. Mange and his colleagues have coined the name 'biodules' for these cells. All the cells are identical. They become differentiated by the job they come to do once the whole array of cells, the organism, has been configured. Just like in a real creature 'only the state of the cell, the combination of the values in its memories, can differentiate it from its neighbours'.[13]

Configuration of the whole array begins with the downloading of the mother cell; typically this is placed into the cell at the bottom left corner. Each of the other cells is then given a location number relative to the mother corner. The configuration program is propagated across the whole array. Each cell knows what to do and which part of its genotype to express thanks to its location on the grid of cells. Again this approach was taken to mimic what happens in living multicellular organisms. Nearly all the cells in an organism contain the entire DNA sequence for that species. A cell knows which part of the genotype to express because of where it is and the job it is doing.

This location information can also help the array cope with faulty or dead cells. If a cell is out of action, the surrounding cells will change their location information to take this into account and start expressing the part of the genotype that is appropriate to the new location they are in.

Mange and his Swiss colleagues have started to create simple self-repairing computer systems using the FPGA arrays. So far these have only been given simple tasks such as checking a string of brackets to ensure that each left bracket is part of a closed pair. Now the research group is starting to turn these simple arrays into more general purpose computers. Recently their work has concentrated on proving that any logic system used in programming can be translated into a genome for an FPGA array. This is important if FPGA-based computers are to be used widely because they will have to perform lots of different tasks. Soon they hope to be testing a living version of the kinds of processors that are found in PCs on desktops today.

Chapter Six

The Machine Stops

But there came a day when, without the slightest warning, without any previous hint of feebleness, the entire communications system broke down, all over the world, and the world as they understood it, ended.

E.M. Forster, *The Machine Stops*[1]

There is nothing special about human beings. The greatest achievement of twentieth-century science has been the removal of humanity from the pedestal upon which it had placed itself. Not everyone accepts this conclusion, even though there is overwhelming evidence to show that at almost every level on which you choose to analyse humans other organisms share many of the same features or do things in the same way.

Artificial life is just one of the fields that is helping us step down from the pedestal by showing us just how much we have in common with every other living thing on this planet and perhaps on other ones too.

The key insight that ALife is helping to establish is that the informational basis of life can be abstracted away from the bodies we find it in and lose nothing in the process. This insight shows that we are right to think that we are more than just the sum of our parts, but it also reveals the slightly less comfortable fact that so is every other living thing on the planet. This is not meant to be the first step in establishing a latter day vitalism because there is nothing mysterious about information or the ways in which the dynamics of information come to dominate. It cannot be distilled and put in a jar, it can only be seen in action. To see it at work you need only look all around you.

ALife, via work with cellular automata and artificial chemistry, is showing that the trends of self-maintenance and reproduction become established at the lowest levels. From this chemical base these trends have managed to create the living world. The implication is that humans are no longer individually special but only as special as everything else around them.

What is important about life, in its real and artificial incarnations, is the organization of the parts not the properties of the parts themselves. Because of this it is possible to study how this organization emerges on computer. Without this convenience the study of life would be a much more difficult enterprise. With it we can start to unravel the conditions under which life arrives and thrives confident that what we are seeing on a computer screen resembles the behaviour of living organisms. Everything that is alive shares these properties.

Some ALife researchers make strong claims about their robotic or software creations. They claim that these creations are not synthetic but in fact are a new form of life that has its own imperatives. If they are correct then this creates a problem. It means that they are engaged in digital naturalism, the study of new species, rather than simply trying to mimic life so that we can better understand the creatures that we share this planet with. So far the ALife movement has not reached a consensus on this point. Some even resist being identified with the ALife movement or deny that there is a movement at all. It is true that the term ALife is preferred more by journalists than by scientists and researchers but it is a useful catch-all phrase for the work that is taking place.

This work by ALife researchers shows that life is a dynamic, self-organizing process that relies on information to keep it intact and developing. Information dynamics comes to dominate at a criticial phase transition – a cusp on either side of which information flows too freely or is frozen and cannot be used. Life strives to keep itself at this point.

ALife research is helping to demystify the world. It shows that life is everywhere and that the forces that shape living things are the same throughout the world. Once we realize that humanity is one point in a landscape of possibilities we also gain insights into other issues, such as consciousness, that previously have seemed intractable.

Although every living thing shares some characteristics there are still widespread differences between separate species. Differences in brain size and the mental life experienced by those creatures with larger brains are very difficult to explain. Clearly humans exist as another point on a continuum of brain functions, but there also seems to be a qualitative difference between the kind of mental experience I possess compared to that of a fly or a dog. Obviously our brains do the same kinds of things, thoughts happen in the same electro-chemical way. There is no need to appeal to quantum effects to explain what is happening in the brain. Also it is clear that intelligent behaviour is not limited to humans. All living things make intelligent choices about what to do, it is only the kinds of things that they think about and the behaviours that it leads to that differ. Even if some of those choices were made many generations ago and now are more reasonably called instincts. Despite this it does seem to be the case that something else is going on inside human brains. But ALife can help here too.

The work of Rodney Brooks and other robot makers is establishing that a lot of our thinking can be explained by its relation to physical activity we have indulged in. Clues to this can be found in language, which is peppered with physical metaphors that make their point by establishing a link with reality and the kinds of things that we do with our bodies. This removes the need for a lot of the baroque mental furniture that AI researchers have previously assumed we needed to get on with our day. If everything relates to or is abstracted from physical behaviour there is no need to rely on symbolic structures or centrally held models of the world to explain thought. Philosophers like P.F. Strawson advanced a similar argument.

This leads to a slightly worrying conclusion that there is no intelligence or knowledge within our brains. All that we are is the sum of our experience and behaviour. What we know is based on what we did and is only meaningful in relation to everything else that we, or other people, have done. We impute intelligence to a creature when we see it moving around purposefully and coping well with the problems that the world throws at it. I suggest that the same is true when we talk of consciousness in humans. It is something that we impute on the basis of behaviour even though we have little or no evidence for believing that it is actually any kind of stuff that can be seen, poked and pointed to.

When asked where and what Cyberspace is science-fiction author Bruce Sterling has said that it is where telephone calls happen. The call does not take place in our heads but in a netherworld between the phones. Maybe the same can be said of consciousness. Perhaps it is not inside anyone's head but takes place between us all. A kind of consensual hallucination that we all happily delude ourselves into believing. This kind of reasoning is open to the accusation that I am not raising the threshold for an artificial lifeform as much as I am lowering the standards by which we judge ourselves.

The philosopher Ludwig Wittgenstein advanced a similar kind of argument. He claimed that when we speak we really do not know what we are talking about. Meanings are warily shared, but I have no way of knowing that what I experience when I see red is what you experience. We simply agree on what we are going to call red. Because of the impossibility of knowing about the inner experience of anyone other than ourselves, Wittgenstein claimed that this led to the inescapable conclusion that we do not know what we are talking about because our language reflects so little of our experience.

Rodney Brooks' work with Cog should go some way to resolving some of these questions and perhaps start to establish that consciousness is not so strange after all. There remains the intriguing possibility that we will be unable to communicate with the sophisticated creations that ALife throws up in the future. Tom Ray has written: 'I believe that a true machine intelligence is likely to be fundamentally different from a human (or any organic) intelligence. A machine intelligence would be even more alien than an intelligence from another planet, because such an extra-terrestrial creature would probably be carbon based, whereas a machine intelligence is not.' Only time and more experiments will tell.

Doom Watch

While ALife is undoubtedly helping to make progress on issues that have taxed thinkers for centuries there are dangers inherent in the work as well. The creation and studying of life is not something that we should do glibly. Just as bringing a child into the world confers considerable responsibilities on the parents of the child, so the creation of new, artificial, life should not be done lightly but with full recognition of what we are attempting.

Chris Langton has warned of some of the dangers. 'There is a caution here that we all must attend to. Attempts to create Artificial Life may be pursued for the highest scientific and intellectual goals, but they may have devastating consequences in the real world, if researchers do not take care to insure that the products of their research cannot "escape", either into computer networks or into the biosphere itself. There is little danger of that at this stage in the field, but the danger is increasing rapidly as we understand more and more about the synthesis of biological phenomena. What will the world be like when the means to produce self-reproducing robots is as widely available as the means to produce self-reproducing computer programs?'[2]

The problem with this approach – for humans at least – is that when an organism or entity is bred for a task its creators do not know how or why it does it. As the intelligence of the organism increases and its behavioural repertoire widens so the ignorance of its creator about its motivations and goals expands. We are close to creating a new race of beings that are as inscrutable as apes and as difficult to communicate with and control. Wittgenstein once said that if lions could talk we would not understand them. Ray, and others, suspect that the same might be true of any ALife entities that learn to communicate.

Using evolution and sexual reproduction to breed better robots – be they built of software or hardware – is producing startling results. Already some telephone calls are being routed using software agents that have been bred for the job and other companies are considering turning over the running of car plants and nuclear power stations to such systems.

What ALife researchers find attractive about this approach is the chance it gives them to play god. They can roll back the evolution of the software agents to an earlier generation if they do not like the way they are going. Doubtless God will be able to fill us in on the problems that arise when you give a creature free will.

The real world is set solid in bones, fur and flesh and this is where software has the edge; it is infinitely more malleable and it reproduces much faster. Recent work on protein electronics promises to make it even more flexible and perhaps even capable of self-reproduction. Soon computers could look very different to their current four-square and functional incarnation. They might exist entirely independently of humans. They may even have to graze rather than plug-in to keep themselves living.

But breeding software to do our bidding is a path fraught with danger. Evolution is all about surprises and unpredictability, forces that are impossible to harness. If we try to shape the offspring of the software we are breeding, we risk reducing adaptability and perhaps stifling the growth of the very organism we are looking for. In fact the reason we are adapting them for our own purposes is to take advantage of these characteristics.

The problem is that once a system is handed over to living, breeding software there is no turning back. The systems many companies are contemplating using this live software to control are so complex and change so fast that it would be impossible to regain control using old methods once the switch is made. So much will have happened during those moments the live software was in charge that it would take weeks of analysis to catch up.

Handing over such a critical system as the telephone network or a power station to a population of living organisms about which we will know progressively less could be the greatest act of betrayal in human history.

There exists the possibility that the software running the network will become more interested in pursuing its own goals rather than those we chose to give it. Certainly early work using sexual software has shown that it tends to spend a lot of time having sex with itself rather than applying itself to the problem it was created to tackle. Unconstrained by morals the software tends also to treat unfit offspring brutally by killing them off if they do not perform as well as expected.

The ethic of evolution is to weed out those who are ageing and obsolete to make way for those better fitted to the world we all find ourselves in. By riding the cusp of chaos in order to breed better software we could be creating our own successors and risking a plunge into anarchy.

I suspect that such doom-laden predictions are a little

overblown. More likely we will learn to live with ALife creations for a long time before we feel any kind of threat from them. The more likely scenario is that we will come to be at ease with robots that help us do our jobs better and view them much as we view microwaves today. They will become familiar tools and possibly companions long before they are smart enough to think about doing away with us. Like children they will need to be taught, but in raising them they will have much to teach us as well.

Notes and References

Introduction

1) Quoted in Yates, F. A. (1964), 248.
2) For an approachable introduction to thermodynamics see Atkins, P. W. (1984).
3) Yates, F. A. (1964) is the best history of the life and philosophy of Giordano Bruno.
4) Enfield is quoted in Brown, S. B., and Wallace, P. M. (1980), 3.
5) *Daily Mail*, 8 August 1996, 3.
6) Travis, J. (1997).
7) Langton, C. G., et al. (1992), 19.
8) Lovelock, J. (1988), 25.
9) Ibid., 24.
10) See Chapter Two, p. 84–90 for a longer discussion of this point.
11) Langton, C. G. (1989), 33.
12) Ibid., 2.

Notes and References

Chapter One: First Stirrings

1) Lovelock, J. (1988), 16.
2) For a readable and up-to-date introduction to the origins of the Universe and astronomy Ferris, T. (1997).
3) The analogy is a little suspect because a journey implies a destination, something that evolution does not admit of. However, the second law of thermodynamics gives history a direction if not an ultimate end point, so the analogy is not wholly miscast.
4) The rhyme review ran as follows:
   Cheap Meat performs passably,
   Quenching the celibates jejeune thirst,
   Portraiture, presented massably,
   Drowning sorrow, oneness cursed.

   The equivalent eras and periods are:
   Cenozoic/Mesozoic/Palaeozoic/Precambrian,
   Quarternary/Tertiary/Cretaceous/Jurassic/Triassic,
   Permian/Pennsylvanian/Mississippian,
   Devonian/Silurian/Ordovician/Cambrian.
   Gould, S. J. (1991), 54.
5) Schopf, J. W. (1993).
6) Mojzsis, S. J., Arrhenius, G., McKeegan, K. D., Harrison, T. M., et al. (1996).
7) Schidlowski, M. (1988).
8) Margulis, L., and Sagan, D. (1986), 71–2.
9) Ibid., 103–4.
10) Fontana, W. (1994), 1.
11) Margulis, L., and Sagan D. (1986), 57.
12) Ibid., 57.

13) Lovelock, J. (1988), 23.
14) Kauffman, S. (1995), 71.
15) McMullin, B. (1997), 38–47.
16) Radetsky, P. (1998), 34–6.
17) Goodwin, B. (1994), 34.
18) Hecht, J. (1997), Lissauer, J. J. (1997), and Shigeru, I., Canup, R. M., and Stewart, G. R. (1997).
19) See p. 33 for more on this point.
20) Deamer, D. W. (1997), 244.
21) McKay, D. S., Gibson, E. K. jr., Thomas-Keprta, K. L., et al. (1996).
22) Radetsky, P. (1998), 37.
23) Horneck, G., Bucker, H., and Reitz, G. (1994).
24) Shock, E. L., and Schulte, M. D. (1990).
25) Radetsky, P. (1998), 39.
26) Zimmer, C. (1995), p. 70.
27) Pargellis, A. (1996a).
28) Ibid., 86–7.
29) See Pargellis, A. (1996b), for details.
30) Margulis, L., and Sagan, D. (1986), 93.
31) Lovelock, J. (1988), 9.
32) Stanley, S. (1988).
33) Gould, S. J. (1991), 55–60.
34) Goodwin, B. (1994), 9–17.
35) Lovelock, J. (1988), 99.
36) Lee, Der-Chuen and Halliday, A. N. (1997).
37) Bakker, R. (1988), 181.
38) Lovelock, J. (1988), 128.

Chapter Two: The Game of Life and How to Play It

1) Heims, S. J. (1980), 38.
2) Ibid., 41.
3) Ibid., 350.
4) Campbell-Kelly, M., and Aspray, W. (1996), 90.
5) Ibid., 94.
6) Poundstone, W. (1985), 18.
7) Pesavento, U. (1995).
8) McMullin, B. (1997).
9) Poundstone, W. (1985), 24.
10) Ibid., 32.
11) Agmon-Snir, Hagai, Carr, C. E. and Rinzel, J. (1998).
12) Levy, S. (1992), 61.
13) Coveney, P., and Highfield, R. (1995), 102.
14) Levy, S. (1992), 68.
15) Wolfram, S. (1984), 419.
16) Langton, C. G., et al. (1992), 46.
17) West, G. B., Brown, J. H., and Enquist, B. J. (1997).
18) Einstein, A. J., Wu, Hai-Shan, and Gil, J. (1998).
19) Agmon-Snir, Hagai, Carr, Catherine E., and Rinzel, J. (1998).
20) Langton, C. G., et al. (1992), 46.
21) Ibid., 84.
22) Ibid., 84.
23) Ibid., 85.
24) Poundstone,W. (1985), 105–6.
25) Adami, C. (1998), 27.
26) Ibid., 41.
27) Simpson, R. (1997).
28) Ibid.

29) For a thorough exploration of Sugarscape see Epstein, J. M., and Axtell, R. (1996).
30) Epstein, J. M., and Axtell, R. (1996), 1.
31) See Glanz, S. (1997), and Orlov, A., et al. (1997).

Chapter Three: Rise of the Robots

1) Rucker, R. (1982), 32.
2) Hodges, A. (1992), 41.
3) Howarth, T. E. B. (1978), 31.
4) Ibid., 92.
5) Ibid., 93.
6) Hobsbawm, E. (1994), 522.
7) Ibid., 523.
8) Hodges, A. (1992), 60.
9) Ibid., 60.
10) Ibid., 94.
11) Ibid., 96.
12) Ibid., 14.
13) Ibid., 107.
14) Brooks, R., and Stein, L. A. (1993), 2.
15) Hodges, A. (1992), 107.
16) Ibid., 130.
17) Ibid., 56.
18) Ibid., 160.
19) Fox, B., and Webb, J. (1997).
20) Crevier, D. (1993), 4.
21) Newquist, H. (1994), 140.
22) Crevier, D. (1993), 114.
23) Allen, P. (1978), 6.
24) Ibid., 8.

25) Symons, J. (1990), 43.
26) Rickelson, J. T. (1995), 91.
27) Philby, K. (1989), 26.
28) Wright, P. (1984), 213.
29) Howarth, T. E. B. (1978), 51.
30) Walter, W. G. (1953), 27.
31) Ibid., 68.
32) Ibid., 112. Norbert Wiener, the pioneer of cybernetics, developed his ideas on feedback while looking for a way to keep artillery locked into a firing position.
33) Ibid., 113.
34) Holland, O. (1997), 34.
35) Ibid., 35.
36) Walter, W. G. (1953), 241.
37) Ibid., 117.
38) Ibid., 112.
39) Only now are robot makers achieving the same sort of sophistication. See the section about robo-biologist Mark Tilden that begins on p. 171.
40) Walter, W. G. (1953), 157.
41) Holland, O. (1997), 34.
42) Walter, W. G. (1953), 118.
43) Brooks, R. (1991), 7.
44) Cooper, R., and Bird, J. (1989), 60 .
45) Levy, S. (1992), 275.
46) Ibid., 275.
47) Ibid., 275.
48) Ibid., 275.
49) Ibid., 275.
50) Churchland, P. S. (1986), 299–312.
51) Crevier, D. (1993), 2, recounts this tale.

52) Churchland, P. S. (1986), 407.

53) Crevier, D. (1993), 1.

54) Kasparov, G. (1996).

55) Brooks, R. (1991), 6.

56) Brooks, R., and Flynn, A. M. (1989), 479.

57) Brooks, R. (1991), 9.

58) Brooks is not alone in thinking such thoughts. See Langton (1989), 38–40.

59) Brooks, R. (1991), 7.

60) Ibid., 15.

61) Brooks, R., and Flynn, A. M. (1989), 479.

62) Brooks, R. (1991), 16.

63) Levy, S. (1992), 280.

64) Langton, C. G. (1989), 11.

65) Brooks, R., and Flynn, A. M. (1989), 481.

66) Ibid., 481.

67) Ibid., 481.

68) Mataric, M (1998), 85–6.

69) Brooks, R., and Flynn, A. M. (1989), 276.

70) Brooks, R., and Stein, L. A. (1993), 3.

71) Ibid., 4.

72) Ibid., 5.

73) Hapgood, F. (1994), 106.

74) Ibid., 107.

75) Smit, M. C., and Tilden, M. W. (1991), 15.

76) Hapgood, F. (1994), 107.

77) Tilden, M. W. (1994), 19.

78) Smit, M. C., and Tilden, M. W. (1991), 16.

79) Ibid., 15.

80) Ibid., 17.

81) Hapgood, F. (1994), 106.

82) Tilden, M. W. (1994), 1.
83) Hasslacher, B., and Tilden, M. W. (1995), 8.
84) Ibid., 5.

Chapter Four: Smaller is Smarter

1) Belt, T. (1874), 26.
2) Holldobler, B., and Wilson, E. O. (1995), 1.
3) Ibid., 7.
4) Ibid., 8–9.
5) Agosti, D., Grimaldi, D., and Carpenter, J. M. (1997).
6) Holldobler, B., and Wilson, E. O. (1994).
7) Collett, T. (1996).
8) Franks, N. (1989), 139.
9) Holldobler, B., and Wilson, E. O. (1994), 107.
10) Minsky, M. (1986).
11) Franks, N. (1989), 139.
12) Belt, T. (1874), 28.
13) Kelly, K. (1994), 366.
14) Ibid., 366.
15) Franks, N. (1989), 142.
16) Ray, T. S. (1991). 'An Approach to the Synthesis of Life'. In Langton, C. G., et al. (1992), 371–2.
17) Levy, S. (1992), 217.
18) Spafford, E. H. (1992), in Langton, C. G., et al. (1992), 727–45.
19) Ray, T. S. (1991), 372.
20) Ibid., 372.
21) Ibid., 375.
22) Ibid., 397.
23) Ibid., 376.

24) Gutowitz, H. (1995).
25) Ray, T. S. (1991), 380.
26) Levy, S. (1992), 221.
27) Coveney, P., and Highfield, R. (1995), 258.
28) Gould, S. J. (1991), for an excellent introduction to the Edicara, 56.
29) Thearling, K., and Ray, T. S. (1994).
30) Taubes, G. (1995).
31) Holland, O. (1997), 186.
32) Ibid., 186–98.
33) Kelly, K. (1994), 111.
34) Bedau, M., Snyder, E., Brown, C. Titus, and Packard, N. (1997), 127.
35) For a thorough explanation of Avida see Adami, C. (1998), 50–53, and Chapter Nine.
36) Bedau, M., Snyder, E., Brown, C. Titus, and Packard, N. (1997), 125.
37) Gould, S. J. (1991), 208.
38) Bedau, M., Snyder, E., Brown, C. Titus, and Packard, N. (1997), 127.
39) *Scientific American.*
40) Taylor, T., and Hallam, J. (1997).
41) Sterling, B. (1992), 15.
42) Appleby, S., and Steward, S. (1994).
43) Harrison, P. F. (1997).
44) Appleby, S., and Steward, S. (1994).
45) See Nwana, H. (1997), for a round-up of current research into agent technology.
46) Appleby, S., and Steward, S. (1994).
47) Schoonderwoerd, R. (1996) report.
48) Dorigo, M., Maniezzo, V., and Colorni, A. (1996).
49) Maes, P. (1994).

50) See the papers by Aimar, M., Arnaud, G., and Dumas, M. (1997), and Rodriguez, J. A., Noriega, P., Sierra, C., and Padget, J., in Crabtree (1997).

Chapter Five: Living Machines

1) Dick, P. K. (1972).
2) Smith, J. M. (1993), 165.
3) Ridley, M (1993), 36.
4) Levy, S. (1992), 157.
5) Ibid., 158.
6) Holland, J. H. (1992), 3.
7) Matthews, R. (1995), 41.
8) Ibid., 43.
9) Ray, T. S. (1994), 220.
10) Ray, T. S. (1994), 221.
11) Thompson, A. (1997), 393.
12) Mange (1996).
13) Ibid., 201.

Chapter Six: The Machine Stops

1) Forster, E. M. (1954), 142.
2) Langton, C. G. (1992), 18, in Langton, C. G., Taylor, C., Farmer, J. D., and Rasmussen, S. (eds.).

Bibliography

Adami, Christoph. 1998. *An Introduction to Artificial Life.* New York: Springer Verlag

Agmon-Snir, Hagai, Carr, Catherine E., and Rinzel, John. 1998. The role of dendrites in auditory coincidence detection. *Nature,* vol. 393, pp. 268–72

Agosti, D., Grimaldi, D., and Carpenter, J. M. 1997. Oldest known ant fossils discovered. *Nature,* vol. 391, 29 January 1997, p. 447

Allen, Peter W. 1978. *The Cambridge Apostles – The early years.* Cambridge: Cambridge University Press

Appleby, Steve and Steward, Simon. 1994. Mobile software agents for control in telecommunications networks. *BT Technology Journal,* vol. 12, no. 2, pp. 104–13.

Atkins, Peter W. 1984. *The Second Law.* New York: W. H. Freeman and Company

Bakker, Robert. 1988. *The Dinosaur Heresies.* London: Penguin

Bedau, Mark, Snyder, Emile, Brown, C. Titus, and Packard, Norman. 1997. A comparison of evolutionary activity in artificial evolving systems and in the biosphere. In *Proceedings of the Fourth European Conference on Artificial Life,* eds. Phil Husbands and Harvey Inman, pp. 125–44. Cambridge, Mass.: MIT Press

Belt, Thomas. 1874. *The Naturalist in Nicaragua*. London: John Murray

Brooks, Rodney, and Flynn, Anita M. 1989. Fast, cheap and out of control: a robot invasion of the solar system. *Journal of the British Interplanetary Society*, no. 42, pp. 478–85

Brooks, Rodney. 1991. Intelligence without reason. MIT AI Memo no. 1293

Brooks, Rodney, and Stein, Lynn Andrea. 1993. Building brains for bodies. MIT AI Memo no. 1439

Brown, S. B., and Wallace, Patricia M. 1980. *Physiological Psychology*. New York: Academic Press

Campbell-Kelly, Martin, and Aspray, William. 1996. *Computer: A history of the information machine*. New York: Basic Books

Churchland, Patricia Smith. 1986. *Neurophilosophy: Toward a unified science of the mind-brain*. Cambridge, Mass.: Bradford Books

Collett, Tom. 1996. Insect navigation en route to the goal: multiple strategies for the use of landmarks. *The Journal of Experimental Biology* 199, pp. 275–35

Cooper, Ray, and Bird, Jonathan. 1998. *The Burden: Fifty Years of Clinical & Experimental Neuroscience*. Bristol: White Tree Books

Coveney, Peter, and Highfield, Roger. 1995. *Frontiers of Complexity*. New York: Ballantine Books

Crabtree, Barry. 1997. *Proceedings of the Second International Conference on the Practical Application of Intelligent Agents and Multi-agent Technology* (PAAM 97). London

Crevier, Daniel. 1993. *AI: The tumultuous history of the search for artificial intelligence*. New York: Basic Books

Deamer, David W. 1997. The first living systems: a bioenergetic perspective. *Microbiology and Molecular Biology Reviews*, vol. 61, no. 2, pp. 239–61

Dick, Philip K. 1972. *The Android and the Human in the Shifting Realities of Philip K. Dick – Selected literary and philosophical writings*, ed. Lawrence Sutin (1995). New York: Vintage

Dorigo, Marco, Maniezzo, Vittorio, and Colorni, Alberto. 1996. The ant system: optimization by a colony of cooperating agents. IEEE transactions on systems, *Man and Cybernetics*–Part b, vol. 26 no. 1, pp. 1–13

Einstein, A. J., Wu, Hai-Shan, and Gil, Joan. 1998. Self-affinity and lacunarity of chromatin texture in benign and malignant breast epithelial cell nuclei. *Physical Review Letters*, vol. 80, no. 2, 12 January 1998, pp. 397–400

Epstein, Joshua M., and Axtell, Robert. 1996. *Growing Artificial Societies: Social science from the bottom-up*. Cambridge, Mass.: MIT Press

Ferris, Timothy. 1997. *The Whole Shebang*. New York: Simon and Schuster

Fontana, Walter. 1994. On Organisation. Talk presented at the Carlo Erba Foundation meeting 'The Future of Science has begun' in Milan

Forster, E. M. 1954. 'The Machine Stops' in *Collected Short Stories*, p. 142. London: Penguin

Fox, B., and Webb, J. 1997. A colossal adventure. *New Scientist*, vol. 154, no. 2081, pp. 38–43

Franks, N. 1989. Army ants: a collective intelligence. *American Scientist*, vol. 77, March–April, pp. 139–45

Glanz, James. 1997. Quantum cells make a bid to outshrink transistors. 277, pp. 898–9

Goodwin, Brian. 1994. *How the Leopard Changed its Spots*. London: Orion Books

Gould, Stephen Jay. 1991. *Wonderful Life*, pp. 53–4. London: Penguin

Gutowitz, Howard. 1994. *Artificial Life Simulators and their Applications*. DRET Technical Report

Hapgood, Fred. 1994. Chaotic robotics. *Wired magazine*. 2.09. pp. 106–7

Harrison, Paul. 1997. Customer service system – past, present, future. *BT Technology Journal*, vol. 15, no. 1, pp 29–45.

Hasslacher, Brosl., and Tilden, M. W. (1995) *Living Machines, Robotics and Autonomous Systems: The biology and technology of intelligent autonomous agents*. LANL Paper LA–UR–94–2636

Hecht, Jeff. The making of a moon. *New Scientist*, vol. 155, no. 2093, p. 8

Heims, Steven J. 1980. *John Von Veumann and Norbert Wiener: From mathematics to the technologies of life and death*. Cambridge, Mass.: MIT Press

Hobsbawm, Eric. 1994. *Age of Extremes*. London: Abacus

Hodges, Andrew. 1992. *Alan Turing: The engima*. London: Vintage

Holland, John Henry. 1992. *Adaptation in Natural and Artificial Systems*. Cambridge, Mass.: MIT Press

Holland, Owen. 1997. Grey Walter: The pioneer of real artificial life. In *Artificial life V: Proceeding of the Fifth International workshop on the synthesis and simulation of Living Systems*. Christopher G. Langton and Katsunori Shimohara, eds., pp. 34–41. Cambridge, Mass.: MIT Press

Holldobler, Bert, and Wilson, Edward O. 1994. *Journey to the Ants*. Cambridge, Mass.: Belknap Press

Horneck, G., Bucker, H., and Reitz, G. 1994. Long-term survival of bacterial spores in space. *Advances in Space Research*, vol. 14, no. 10, pp. 41–5

Howarth, T. E. B. 1978. *Cambridge Between Two Wars.* London: Collins

Husbands, Phil and Inman Harvey, eds. 1997. *Proceedings of the Fourth European Conference on Artificial Life.* Cambridge, Mass.: MIT Press

Kasparov, Garry. 1996. The day that I sensed a new kind of intelligence. *Time,* vol. 147, no. 13, p. 47

Kauffman, Stuart. 1995. *At Home in the Universe.* Oxford: Oxford University Press

Kelly, Kevin. 1994. *Out of Control: The new biology of machines.* London: Fourth Estate

Langton, Christopher G. 1986. Studying artificial life with cellular automata. Physica D. 22. pp. 120–49

Langton, Christopher G. 1989. Artificial life. In *Artificial Life,* ed. Christopher Langton, pp. 1–47. Redwood City, CA: Addison Wesley

Langton, Christopher G., Taylor, C., Doyne Farmer, J., and Rasmussen, S., eds. 1992. *Artificial Life II: Proceedings of an interdisciplinary workshop on the synthesis and simulation of living systems.* Redwood City, CA: Addison-Wesley

Langton, Christopher G., and Shimohara, Katsunori. 1997. *Artificial Life V: Proceedings of the fifth international workshop on the synthesis and simulation of living systems.* Cambridge, Mass.: MIT Press

Lee, Der-Chuen, and Halliday, A. N. 1997. Core formation on Mars and differentiated asteroids. *Nature,* vol. 388, pp. 854–7

Levy, Steven. 1992. *Artificial Life: The quest for a new creation.* London: Penguin

Lissauer, J. J. It's not easy to make the moon. *Nature,* vol. 389, 25 September 1997, pp. 327–8

Bibliography

Lovelock, James. 1988. *The Ages of Gaia.* Oxford: Oxford University Press

Maes, P. Agents that reduce work and information overload. *Communications of the ACM,* vol. 37, no. 7, July 1994, pp. 31–40.

Mange, D., Goelle, M., Madon, D., Staliffer, A., Tempesti, G., Durand, S. 1996. Embryonics: a new family of coarse-grained field-programmable gate array with self-repair and self-reproducing properties. In *Towards Evolvable Hardware, Lecture Notes in Computer Science,* pp. 197–220. Berlin: Springer Verlag

Mataric, Maja. 1998. Behaviour-based robotics as a tool for synthesis of artificial behaviour and analysis of natural behaviour. *Trends in Cognitive Science,* 2(3), March 1998, pp. 82–7

Matthews, Robert 1995. Hard maths? No problem: how do you work out the answers to questions that would take a desktop computer fifteen million years to solve. *New Scientist,* vol. 148, no. 2001, p. 41

Margulis, Lynn, and Sagan, Dorian. 1986. *Microcosmos.* Berkeley: University of California Press

McKay, D. S., Gibson, E. K. jr., Thomas-Keprta, K. L., et al. 1996. Search for past life on Mars: possible relic biogenic activity in Martian meteorite ALH84001. *Science,* vol. 273, pp. 924–30

McMullin, Barry. 1997. Rediscovering computational autopoiesis. In *Proceedings of the Fourth European Conference on Artificial Life,* eds. Phil Husbands and Harvey Inman, pp. 38–47. Cambridge, Mass.: MIT Press

Minsky, Marvin. 1986. *The Society of Mind.* New York: Simon and Schuster

Mojzsis, S. J., Arrhenius, G., McKeegan, K. D., Harrison, T.

M., Nutman, A. P., and Friends, C. R. L. 1996. Evidence for life on Earth before 3,800 million years ago. *Nature*, vol. 384, pp. 55–9

Newquist, Harvey. 1994. *The Brain Makers*. Indiana: Sams Publishing

Nwana, Hyacinth S. 1997. *Software Agents & Soft Computing: Towards Enhancing Machine Intelligence*. Berlin: Springer Verlag

Orlov, A. O., Amlani, I., Bernstein, G. H., Lent, C. S., and Snider, G. L. 1997. Realization of a functional cell for quantum-dot cellular automata. *Science*, vol. 277, pp. 928–30

Pargellis, Andrew. 1996a. The spontaneous generation of digital 'Life'. Physica D. 91, pp. 86–96

Pargellis, Andrew. 1996b. The evolution of self-replicating computer organisms. Physica D. 98, pp. 111–27

Pesavento, U. 1995. An implementation of Von Neumann's self-reproducing machine. *Artificial Life*, vol. 2, no. 4, summer 1995, pp. 337–54

Philby, Kim. 1989. *My Silent War*. London: Grafton

Poundstone, William. 1985. *The Recursive Universe*. Oxford: Oxford University Press

Radetsky, Peter. Life's crucible. *Earth*, vol. 7, no. 1, February 1998, p. 39

Ray, T. S. 1991. An approach to the synthesis of life. In *Artificial Life II*, eds. C. Langton, C. Taylor, J. D. Farmer and S. Rasmussen. Santa Fe Institute Studies in the Sciences of Complexity, vol. XI, pp. 371–408. Redwood City, CA: Addison-Wesley

Ray, T. S. 1994. An evolutionary approach to synthetic biology: zen and the art of creating life. *Artificial Life 1* (1/2), pp. 195–226. Cambridge, Mass.: MIT Press

Bibliography

Richelson, Jeffrey, T. 1995. *A Century of Spies*. Oxford: Oxford University Press

Ridley, Matt. 1993. Is sex good for anything? *New Scientist*, vol. 140, no. 1902, p. 36

Rucker, Rudy. 1982. *Software*. London: Penguin

Schidlowski, Manfred. 1988. A 3800 million year isotopic record of life from carbon in sedimentary rocks. *Nature*, vol. 333, pp. 313–18

Schoonderwoerd, Ruud. 1996. Collective intelligence for network control. PhD thesis

Schopf, J. William. 1993. Microfossils of the early archaean apex chert: new evidence of the antiquity of life. *Science*, vol. 260, pp. 640–6

Shigeru, I., Canup, R. M., and Stewart, G. R. 1997. Lunar accretion from an impact-generated disk. *Nature*, vol. 389, 25 September 1997, pp. 353–7

Shock, E. L., and Schulte, M. D. 1990. Amino acid synthesis in carbonaceous meteorites by aqueous alteration of polycyclic aromatic hydrocarbons. *Nature*, vol. 343, pp. 728–31

Simpson, Roderick. 1997. The brain builder. *Wired* magazine, 5 December 1997, pp. 234–5

Smit, M. C., and Tilden, M. W. 1991. Living Machines. *Algorithm 2.2*, March 1991, pp. 15–19

Smith, John Maynard. 1993. *Did Darwin Get It Right? Essays on games, sex and evolution*. p. 165. London: Penguin.

Spafford, Eugene, H. 1992. Computer viruses – a form of artificial life? In *Artificial Life II*, eds. Christopher G. Langton, Charles Taylor, J. Doyne Farmer and Steen Rasmussen, pp. 727–45

Stanley, Steven 1988. Palaeozoic mass extinctions: shared patterns suggest global cooling as a common cause. *American Journal of Science*, 288, pp. 334–52

Sterling, Bruce. 1992. *The Hacker Crackdown*. London: Penguin

Symons, Julian. 1990. *The Thirties and the Nineties*. Carcanet: Manchester

Taubes, Gary. The rise and fall of thinking machines, *Inc.* magazine, 1995, pp. 61–7

Taylor, Tim, and Hallam, John. 1997. Studying evolution with self-replicating computer programs. In *Proceedings of the Fourth European Conference on Artificial Life*, eds. Phil Husbands and Harvey Inman, pp. 550–9. Cambridge, Mass.: MIT Press

Tilden, M. W. 1994. *Beam Robotics V4.0 Event Rules and General Guidelines*, p19. New Mexico: Beam Robotics

Thearling, Kurt, and Ray, T. S. 1994. Evolving multi-cellular artificial life. In *Artificial Life IV Conference Proceedings*, eds. Rodney A. Brooks and Pattie Maes, pp. 283–8. Cambridge, Mass.: MIT Press

Thompson, Adrian. 1997. An evolved circuit, intrinsic in silicon, entwined with physics. In *Proceedings of the First International Conference on Evolvable Systems: From Biology to Hardware* (ICES 96), eds. Higuch, T., and Iwata, M., pp. 390–405. Berlin: Springer Verlag LNCS 1259

Travis, John. 1997. Eye-opening gene. *Science News*, 151. pp. 288–9

Walter, William Grey. 1953. *The Living Brain*. Penguin: London

Walter, William Grey. 1956. *Further Outlook*. London: Duckworth and Co.

West, G. B., Brown, J. H., and Enquist, B. J. 1997. A general model for the origin of allometric scaling laws in biology. *Science*, vol. 276, 4 April 1997, pp. 122–6

Willmer, P. G., and Stone, G. N. 1997. How aggressive ant-guards assist seed-set in acacia flowers. *Nature*, vol. 388, pp. 165–7

Wolfram, S. 1984. Cellular automata as models of complexity. *Nature*, vol. 311, no. 4, October 1984, pp. 419–24

Wright, Peter. 1984. *Spycatcher*. London: William Heinemann

Yates, Frances A. 1964. *Giordano Bruno and the Hermetic Tradition*. Chicago: University of Chicago Press

Zeyl, Clifford, and Bell, Graham. 1997. The advantage of sex in evolving yeast populations. *Nature*, vol. 388, pp. 465–8

Zimmer, Carl. First cell. *Discover*, November 1995, pp. 70–78

Index

Index

Bergson, Henri 4
Big Bang, theory of 17–18
biodules 269–70
biology 15, 18–19, 92–3
biotic organisms 41
Blunt, Anthony 134, 135
Bombe machine 121–2
Boolean logic 105
Born, Max 110
brain, artificial, construction of 93–5
Braitenberg, Valentino 142
BRL *see* Ballistics Research Laboratory
Brooks, Rodney 142–6, 152–71, 186, 230, 275–6
Brown, James 82–3
Bruno, Giordano 1, 4–5
Bruten, Janet 232, 235
bubbles, research into 35–8
Bugs (life-simulation computer program) 216–7
Burgess, Guy 132, 134–5
Burks, Arthur 61, 68, 77, 92, 250
Buss, Leo 28

calcium 50
CAM-8 78
Cambrian period 46, 202–3, 212, 216, 221, 260
cancer, use of fractal patterns in diagnosis 83–4
carbon 30–31; organic 37
carbon dioxide 46, 49, 52
Carroll, Lewis (*ps.* Charles Dodgson), *Alice Through the Looking Glass* 244
CAs *see* cellular automata
cell structure, early 36–7, 38
cellular automata (CAs) 70, 76–90, 95–6, 103, 158; *see also* quantum cellular automata
chaos mathematics 12
chaos theory 12–14, 86
Church, Alonzo 118
Churchill, Sir Winston 122
Cochrane, Peter 229, 232, 245–7
Codd, Ted 92

Cog (robot) 167–71, 276
Coker, Paul 257
Colossus machine 122–3
complexity theory 47
computers, invention of 60–64
Connell, Jonathan 145, 159, 162
consciousness 5
Conway, John Horton 70, 80, 85, 91, 95
Cooper, Ray 143
Copernicus, Nicholas 4
Cosmos (life-simulation computer program) 218–23
Craik, Kenneth 137
CSS *see* Customer Service System
Customer Service System (CSS) 227–9
cyanobacteria 45
cytosine 35

Daily Mail (newspaper) 6
Darwin, Charles 24, 73
Dawkins, Richard: *The Blind Watchmaker* 6; *The Selfish Gene* 47
de Garis, Hugo 93–5
Deamer, David 35–8, 39
Dean, Jeff 101
Deep Blue (computer) 151–2
Descartes, René 147–8; *Meditations* 147
di Caro, Gianni 235
Dick, Philip K., *The Android and the Human* 240
dinosaurs 51
DNA 8–9, 29–31, 35–8, 43, 86, 200–201, 207, 222–3, 270
Dorigo, Marco 235, 236–7, 239, 254
Dovey, Vivian *see* Walter, Vivian
Drake, Sir Francis 73
Drescher, Melvin 213
Driesch, Hans 4
dualism, Aristotelian 5

earth: age of 19; beginnings of life on 40; atmosphere of 49–50

Index

Index

Index

Index